Elementary School Mathematics for Parents and Teachers

Volume 1

Elementary School Mathematics for Parents and Teachers

Volume 1

Raz Kupferman
The Hebrew University, Israel

NEW JERSEY · LONDON · SINGAPORE · BEIJING · SHANGHAI · HONG KONG · TAIPEI · CHENNAI · TOKYO

Published by

World Scientific Publishing Co. Pte. Ltd.

5 Toh Tuck Link, Singapore 596224

USA office: 27 Warren Street, Suite 401-402, Hackensack, NJ 07601

UK office: 57 Shelton Street, Covent Garden, London WC2H 9HE

Library of Congress Cataloging-in-Publication Data
Names: Kupferman, Raz.
Title: Elementary school mathematics for parents and teachers / by Raz Kupferman
   (The Hebrew University, Israel).
Description: New Jersey : World Scientific, 2015– | Includes index.
Identifiers: LCCN 2015030122 | ISBN 9789814699907 (hardcover : alk. paper : v. 1) |
   ISBN 9789814699914 (pbk. : alk. paper : v. 1)
Subjects: LCSH: Education--Parent participation. | Mathematics--Study and teaching (Elementary)
Classification: LCC QA135.6 .K825 2015 | DDC 372.7--dc23
LC record available at http://lccn.loc.gov/2015030122

British Library Cataloguing-in-Publication Data
A catalogue record for this book is available from the British Library.

Illustrations by Philip Pekar

Copyright © 2016 by World Scientific Publishing Co. Pte. Ltd.

All rights reserved. This book, or parts thereof, may not be reproduced in any form or by any means, electronic or mechanical, including photocopying, recording or any information storage and retrieval system now known or to be invented, without written permission from the publisher.

For photocopying of material in this volume, please pay a copying fee through the Copyright Clearance Center, Inc., 222 Rosewood Drive, Danvers, MA 01923, USA. In this case permission to photocopy is not required from the publisher.

Preface

How did I get to write this book?

I am a professor of mathematics at the Hebrew University of Jerusalem in Israel. In 2008, I inaugurated, together with a group of vigorous graduate students, a professional development program (PD) in mathematics for elementary school teachers. The purpose of our program was to enhance the content knowledge (CK) of elementary school teachers in Israel, as most of them lack post-scholar mathematical education. The uniqueness of our PD was that all the instructors were active mathematicians. This created a fascinating encounter between two groups of people—mathematicians and elementary school teachers—who have very different perspectives on what mathematics is and how it should be taught.

One of my first revelations, when we started to plan the PD, was that there is not a single mathematics textbook (in Hebrew) for teachers. School teachers don't have a single resource to compensate for their lack of content knowledge. This applies also to resources of pedagogical content knowledge (PCK), a term coined by Shulman for content knowledge that is particularly relevant for teachers.

This lack of resources is particularly startling from the perspective of a university professor. I cannot imagine my work as a lecturer without an abounding literature, comprising hundreds of textbooks and reference books on any subject I might ever teach. I benefit from such dedicated literature in all the stages of my teaching: to consolidate my own knowledge, to build a syllabus, to construct lesson plans, to gain exposure to the perspectives of other lecturers, and to glean plenty of examples and exercises. Such literature, as I discovered, is not available to the thousands of mathematics teachers in Israel.

While researchers unanimously agree on the strong correlation between CK and high quality teaching, the very idea that teachers should gain expertise in elementary school mathematics is foreign to the general public. Everybody understands that teachers need to be trained in order to pass their knowledge on to their students, but only a few think that teachers need to consolidate their knowledge of the subject matter. Elementary school mathematics is usually perceived as rudimentary common knowledge. "Aren't teachers proficient in the four operations of arithmetic?", I was asked more than once, or, "Can't teachers distinguish a triangle from a square?".

CK and PCK involve much more than "distinguishing a triangle from a square". Being acquainted with the curriculum is certainly a prerequisite for teachers, but the more important factor is the depth of their knowledge. Mathematics is not just an assortment of facts that need to be known and skills that need to be practiced. Mathematics is perhaps the climax of human intelligence: a science in which abstraction and generalization are central concepts; a science characterized by recurring patterns and inter-relations; a science that offers an inexhaustible collection of fascinating riddles for all ages and levels.

Good teachers must have an in-depth knowledge of mathematics in order to pass on the essence of this science and the joy of its practice to their students. High quality teaching requires the ability to distinguish between the essential and the inessential, the general and the specific, as well as to be able explain a concept in several ways, to find the most powerful metaphors and forms of representation, to provoke critical thinking and self-analysis, to choose good examples and exercises, and to rapidly and accurately identify students' errors and misconceptions. These diverse abilities don't appear out of the blue. They have to be acquired and honed. One way of doing this is through dedicated professional literature. While I was preparing my own lesson plans for the PD, I started to write down some notes on elementary school mathematics, which, I hoped, would be valuable for teachers. Three years later, I decided to collect this assortment of lecture notes and lesson plans into a textbook on elementary school mathematics.

Who should read this book?

When I started to write this book, my target audience was elementary school teachers. In the course of writing I became aware of a much larger group of people who could benefit from this book—parents of elementary school children. Parental involvement in their children's education has been on the rise for many years: at home, whether through home schooling or in public education, through PTA activity, and more. With this in mind, I decided to address this book to everyone involved in children's mathematical education, whether as a certified teacher or as a parent. Henceforth, whenever I refer to **teachers** or **educators**, these should be interpreted in a wide sense: every parent is automatically a teacher!

A third category of target readers comprises anyone who wishes to rediscover the mathematics they studied at a young age from a more mature point of view. In summary: this is a book for adults on mathematics for children.

The benefits that readers are expected to gain are the following:

> (1) **Content knowledge**: This book is foremost a treatise on elementary school mathematics. It exposes the reader to the full school curriculum in a logical and systematic order. It provides the reader with a coherent exposition that

integrates between the various topics, thus presenting mathematics as a whole. This unified view, combined with a meticulous attention to details, provides educators with the depth of knowledge required by their profession.

(2) **Pedagogic content knowledge**: While PCK is commonly contrasted to "pure" CK, the distinction between the two is often fuzzy. For example, the order of exposition, and the choices made all along this book as to which topic to expand on versus which topics to reduce, constitute important PCK.

In addition, this book offers the reader traditional PCK: Facts and properties are exposed and explained in more than one way; definitions, theorems and proofs are adapted to fit early stages of math education; distinctions between "easy" and "hard" are stressed and analyzed; the interplay between the concrete and the abstract is emphasized; recurring errors and suggestions on how to eradicate them are presented; relations between the current topic and previously studied topics as well as material to be taught are revealed, thus seeding the foundations for future learning; instructive metaphors and useful means of representations are incorporated in every subject; examples, exercises, and enjoyable activities are advocated.

(3) **Get a glimpse on how mathematicians think**: While some may wonder whether this is a blessing or a curse, I believe that educators have much to benefit from a mathematician's exposition of elementary school mathematics. I have always believed the distinction between elementary and advanced mathematics to be artificial and inappropriate. In a sense, the more (seemingly) elementary mathematics is, the closer it is to the most basic human ways of reasoning, abstracting, and generalizing. These three skills are the crux of a mathematician's work, and they constitute the main objective of math education.

(4) **Activities**: CK and PCK are for educators. Children, especially young children, learn through concrete riddles and problems. It is through repeated experience that their personal perception of mathematics forms. Throughout this book, I have scattered explicit suggestions for activities or games that can be played either at home or in class.

(5) **Fun**: As the reader must have guessed by now, I view mathematics as a pinnacle of human inventions. While mathematics is usually advocated for its usefulness, its beauty is, in my opinion, downplayed. I believe this book can provide many adults with the corrective experience that they deserve. Those, who as children, strived to survive mathematics at last have the opportunity to enjoy it.

What is in this book?

This book is the first volume out of two. The two volumes cover the elementary school math curriculum common in most parts of the world. The first volume focuses on the early years (roughly Grades K-2). The second volume focuses on the subsequent years (roughly Grades 3-6); this classification into grade levels should only be viewed as a rough estimate, not only because curricula vary between countries, but mainly because the same subject may be taught in different grades, at an increasingly advanced level of sophistication.

The present volume is divided into two main parts:

(a) Chapters 1–12 are concerned with numbers and their representation, and with the four operations of arithmetic.

Chapter 1 starts with an exposition of the number concept, and its uses in expressing quantities. Chapter 2 is concerned with our decimal numeral system. Chapter 3 presents an overview on the four operations of arithmetic. The following chapters describe each operation separately, notably its interpretations and properties (addition in Chapter 4, subtraction in Chapter 5, multiplication in Chapter 7, and division in Chapter 8). Chapter 6, as an interlude, presents the concept of even and odd numbers. At this stage we only explore the evaluation of arithmetic operations within the numerical ranges required in the lower grades. Chapter 9 revisits the decimal numeral system, this time at the level needed to understand the principles underlying computational algorithms (addition in Chapter 10 and subtraction in Chapter 11). Chapter 12 presents an extra-curricular topic: base-5 arithmetic. It uses a technique called **estrangement**, which means removal from an accustomed place. The purpose of this chapter is to provide us with more insights into our own decimal system, and give us a better feeling for the difficulties experienced by young school children.

(b) Chapters 13–18 are concerned with geometry. Chapter 13 lays general foundations, relevant to all the following chapters. Chapter 14 deals with lines and segments. The focus of Chapter 15 is length and its measurement. Chapter 16 deals with angles and their measurement. Chapter 17 presents polygons, their classification, and the distinction between their defining and derived properties. Chapter 18 deals with the notion of area and its measurement.

Volume 2 contains the following topics: computation algorithms of multiplication and division; the syntax of arithmetic expressions, commonly known as the order of operations; division with remainders; the concepts of divisibility and primality; divisibility signs; fractions, their interpretation and their incorporation within the number system; fraction comparison; arithmetic operations with fractions; decimal versus rational representations of fractions; percentages and their uses; concepts in statistical data; circles and discs; volume and its measurement.

How to read this book?

I wrote this book with the vision of its readers reading it back to back. This vision conforms to my perception of mathematics being a whole. Yet, this whole has a structure, and the ability to partition this whole into definite sub-topics, each bearing its own title, is an important component of PCK. This structure is reflected in the division of the book into focused chapters, which has the added advantage of inviting educators to read isolated chapters before they start teaching a new topic.

A mathematical treatise is not a novel. Even though I did my best to turn this book into a reader-friendly text, it is very likely that readers might have to stop at times and think. In fact, I would be concerned if this weren't the case. Don't expect to gain new insights without sweating from time to time.

As a result of the many inter-correlations between the various topics, I have included many cross-references between the different chapters. While following cross-references has the obvious disadvantage of spoiling the flow of the reading, bear in mind that the understanding of those inter-relations is an integral part in the understanding of a subject.

As advertised, some sections expose the reader with material that is either extra-curricular, or, taught later in the course of mathematical education. These sections can be skipped at a first reading, but keep in mind that they contain some of the most beautiful and mind-broadening topics in this book.

Every chapter ends with a set of exercises. These exercises are multi-purpose: some exemplify topics learned in the chapter; some challenge the reader with teaching dilemmas; some present typical errors made by children, and ask the reader to offer ways of intervention. Remember that this book is addressed to adults, and so are also the exercise sets. Yet, some of the exercises can serve, perhaps after some adaptation, as challenges for advanced children.

The book has two indices. The first is a "standard" index—an alphabetic list of keywords. The second index is sorted by tags from the American Common Core Standard.

Pedagogical principles

In a textbook of this type, there can be no sharp distinction between mathematics and pedagogy. Even though this is a book for adults, its writing was strongly influenced by my perception on how mathematics ought to be presented to children. In this section I spell out some of the principles I believe should prevail in math education, especially for young children. These principles should serve as guidelines whenever considering transferring contents from this book to children.

(1) **Unpacking**: One of most important challenges in math teaching is to decompose mathematical concepts into elementary constituents. Educators must be aware of all, sometimes subtle, dependencies between various topics. If Subject A is built upon Subjects B and C, then B and C have to be taught before A, and in a manner that paves the way to the later assimilation of A. For example, arithmetic relies on the number concept and counting. Teachers of the lower grades must have this in mind at all times, and make sure that their students internalize all aspects of numbers and counting that are relevant to arithmetic at the proper time.

(2) **Concrete and abstract**: Mathematics is the art of abstraction, but abstraction is not a topic in mathematics; it is a process that occurs in an individual's mind. Moreover, any abstraction must relate to something—to the concrete entities it is an abstraction of. Even though the ultimate goal of math teaching is to endow the learner with the powers of abstract reasoning, abstraction cannot be taught per se. Most of the learning process must occur on the concrete level. It is through a profusion of concrete mathematical experiences, that children gradually develop the ability to draw general conclusions. Please note that something abstract for one person may well be concrete for another. Manipulating numbers is abstract for a kindergartner, who would benefit more from manipulating and counting material objects, while it is concrete to a highschooler, compared to manipulating algebraic expression.

(3) **Learning by doing**: "I hear and I forget; I see and I remember; I do and I understand" is a proverb attributed to Confucius. This short adage encapsulates an uncontested truth: active learning is way more effective than passive learning. When it comes to children (in fact, also when it comes to adults), embarking into an activity is called playing, and luckily, children love playing. This presents educators with the challenge of creating engaging, enjoyable activities, aimed at attaining specific pedagogic goals.

(4) **No nonsense**: Most textbooks and computer applications tend to sugarcoat their content, believing that it makes math more attractive to children. Thus, children find themselves playing with number-eating cows, popping clouds labeled by numbers, and so on. The problem with sugarcoating is that it conveys the message that math is not sufficiently attractive in its own right. This is a self-fulfilling prophecy. Math can be made fascinating without decorating it with infantile metaphors. While the boundaries between the playful and the infantile are sometimes fuzzy, and are a matter of taste, there is one thing I am sure of: if educators don't find the mathematical content interesting, this message will be transferred to children.

(5) **Challenging**: One of the disagreeable realizations I came up with through my encounter with teachers, was their fear of challenging children. "We should not confuse our students" is a warning I have heard too often. The teachers'

fears, of course, do not come out of the blue. They are deep rooted in a general atmosphere that tends to over-protect children from failure.

To me, math without initial confusion is an oxymoron. The whole essence of math is to disentangle mind-boggling riddles with the power of our brain. Moreover, I don't believe math can be fun without a challenge. Children don't buy the "good job" if they feel that the task they have just fulfilled requires no effort. What needs to be changed is the attitude toward errors. Errors do not mean failure; errors are consequences of trying and not giving up. There can be no meaningful math without a univocal legitimation to make errors.

Matific

In 2012, I co-founded Slate Science, the creator of Matific—a suite of educational apps for children in Grades K to 6. Matific was designed and implemented by a team of scientists, teachers and software experts who share a common passion for endearing math on children. It comprises many hundreds of atomic units called **episodes**, each offering a progression of hands-on activities that address one or more mathematical concepts, skill or insight. The development of Matific episodes is guided by the pedagogical principles listed above.

Acknowledgments

I am indebted to a great number of people, who helped me shape this book since its very early drafts. Foremost, I would like to thank the wonderful team of young mathematicians who joined me in inaugurating and carrying out the PD: Noa Eidelstein, Menny Aka, Shaull Almagor, Ayala Byron, Chaim Even Zohar, Ariel Goldstein, Yannai Gonczarowski, Shlomi Hod, Or Landsberg, Noa Nitzan, Ori Parzanchevsky, Shir Peled, Alon Pinto, Doron Puder, Shani Regev, Ori Rosenstein, Ron Rosenthal, Moria Sigron, and Amitai Zernik. We were all closely monitored by Jason Copper and mentored by Abraham Arcavi, who was the one to first suggest the writing of this book.

The PD would have not become a reality without the massive cooperation of the Israeli Ministry of Education. For this I am indebted to the Superintendent of Mathematics Education, Hannah Perl, the national counselor Tami Giron, and the municipal counselors, Nava Atad, Cilla Rott and Tami Razvag.

I am happy to mention two books that served as inspiration: *A Third Divided by a Quarter* by Mira Ofran, and *Arithmetic for Parents* by Ron Aharoni (both published in Hebrew).

It is hard to imagine this book without the contribution of those who have read it, and provided me with invaluable comments, criticism, and constructive suggestions: Abraham Arcavi, Itamar Cwik, Rachel Deitcher, Lisa Feldman, Orna Kupferman, Eran London, Ron Rosenthal, Shimon Schocken, and Yossi Shamai.

This book was originally written in Hebrew. Gal Kusturica was the editor of the Hebrew edition, to whom I owe the transformation of this book from a pile of notes to a (hopefully) coherent and accessible textbook. I am grateful to Ella Jacoby-Bashan for the English translation, and to Sharon Hirsch for her meticulous copy-editing.

Illustrations and graphics play a very important role in books of that type. I am grateful to Philip Pekar for the illustrations, and to Alex Bernstein for useful graphic advice.

Finally, I would like to express my appreciation to the Hebrew University, for its commitment to education, and for granting us all with the academic freedom to apply our knowledge wherever we believe it can be useful.

Contents

Preface   v

Acknowledgments   xiii

1. Counting and Natural Numbers   1
   - 1.1 The natural numbers 1
     - 1.1.1 Development of the number concept 1
     - 1.1.2 Matching between sets 2
     - 1.1.3 Counting 3
     - 1.1.4 The number zero 5
   - 1.2 Order relations between numbers 6
     - 1.2.1 Comparison between sets 6
     - 1.2.2 Comparison between numbers 7
     - 1.2.3 Successor and predecessor 8
     - 1.2.4 There are infinitely many natural numbers 9
     - 1.2.5 The number line diagram 9
   - 1.3 Natural numbers as determining order 10
   - 1.4 Estimation 10
   - 1.5 More on numbers and sets 11
     - 1.5.1 What is *three*? 11
     - 1.5.2 Comparison of infinite sets 12

2. The Decimal Representation System   15
   - 2.1 Development of the decimal system 15
     - 2.1.1 The legend of the shepherd 15
     - 2.1.2 Ancient decimal representation 16
     - 2.1.3 Place-value notation 18
     - 2.1.4 Zero as a place holder 18
   - 2.2 Early age learning of the decimal system 20
     - 2.2.1 The hundred table 21
   - 2.3 Numerals in other cultures 22
     - 2.3.1 Ancient Egypt 22

	2.3.2	China	23
	2.3.3	Ancient Rome	23

3. The Four Operations of Arithmetic   25

- 3.1 Arithmetic operations 25
  - 3.1.1 Operators and operands 26
  - 3.1.2 Concatenating operations 26
- 3.2 Arithmetic expressions 27
- 3.3 Teaching the operations of arithmetic 28

4. Addition   31

- 4.1 What is addition? 31
  - 4.1.1 Addition as a model for joining together 31
  - 4.1.2 Addition as a model for appending 32
  - 4.1.3 The addition operation 32
- 4.2 Word problems 34
  - 4.2.1 Modeling with addition 34
  - 4.2.2 Interpretation 34
- 4.3 The properties of addition 34
  - 4.3.1 The commutative property 35
  - 4.3.2 The associative property 36
  - 4.3.3 The identity property of zero 39
  - 4.3.4 Laws of variation 40
- 4.4 Evaluating addition 41
  - 4.4.1 If you can count, you can add 41
  - 4.4.2 Addends whose sum is 10 43
  - 4.4.3 Addition within 20 43
  - 4.4.4 Using the properties of addition 45
  - 4.4.5 The addition table 45
  - 4.4.6 Addition on the number line 46

5. Subtraction   49

- 5.1 What is subtraction? 49
  - 5.1.1 Subtraction as a model for taking away 50
  - 5.1.2 Subtraction as a model for taking apart 51
  - 5.1.3 Subtraction as a model for comparison 52
  - 5.1.4 Subtraction as a model for complementation 52
  - 5.1.5 The subtraction operation 53
  - 5.1.6 Subtraction on the number line 54
- 5.2 Word problems 55
  - 5.2.1 Modeling with subtraction 55
  - 5.2.2 Interpretation 56
- 5.3 The properties of subtraction 57
  - 5.3.1 Subtraction is not commutative 57

		5.3.2	Subtraction is not associative	58

- 5.3.2 Subtraction is not associative 58
- 5.3.3 Laws of variation 58
- 5.3.4 Adjoint subtraction equation 62
- 5.3.5 Subtraction of zero 63
- 5.4 Evaluating subtraction 63
  - 5.4.1 If you can count, you can subtract 63
  - 5.4.2 Backward and forward counting 63
  - 5.4.3 Subtraction within 20 65
  - 5.4.4 Validation 66
- 5.5 Negative numbers 66

6. Even and Odd Numbers — 69

- 6.1 Definition of parity 69
- 6.2 Arithmetic of parity 72
- 6.3 Determining the parity of a number 74
- 6.4 More on parity 75
  - 6.4.1 Why don't we define parity for fractions? 75
  - 6.4.2 Parity in error control 76

7. Multiplication — 79

- 7.1 What is multiplication? 79
  - 7.1.1 Multiplication as a model for repeated addition . 79
  - 7.1.2 Multiplication as a model for proportional comparison . 80
  - 7.1.3 The multiplication operation 80
  - 7.1.4 Multiplication and counting 81
  - 7.1.5 Rectangular arrays 82
- 7.2 Properties of multiplication 82
  - 7.2.1 The commutative property 82
  - 7.2.2 The associative property 84
  - 7.2.3 The distributive property 86
  - 7.2.4 Laws of variation 87
  - 7.2.5 The identity property of one 87
  - 7.2.6 Multiplication by zero 87
- 7.3 Word problems 88
  - 7.3.1 Repeated addition 88
  - 7.3.2 Combinatorial problems 88
- 7.4 Evaluating multiplication 91
  - 7.4.1 If you can count, you can multiply 91
  - 7.4.2 The multiplication table 92
  - 7.4.3 Various evaluation strategies 92
  - 7.4.4 Multiplication by 10 94

8. Division — 97

- 8.1 What is division? 97

	8.1.1	Division as a model for sharing	97
	8.1.2	Division as a model for rationing	98
	8.1.3	The division operation	99
8.2	Word problems		100
	8.2.1	Modeling with division	100
	8.2.2	Interpretation	101
	8.2.3	Sneak preview: Fraction division	102
8.3	Properties of division		103
	8.3.1	Division is not commutative	103
	8.3.2	Division is not associative	103
	8.3.3	The distributive property	104
	8.3.4	Laws of variation	106
	8.3.5	Interchanging multiplication and division	108
	8.3.6	Adjoint division equation	108
	8.3.7	Division of zero and division by zero	109
	8.3.8	Division by 1	112
	8.3.9	Division of a number by itself	112
8.4	Evaluating division		113
	8.4.1	If you can count, you can divide	113
	8.4.2	Evaluation by repeated addition	113
	8.4.3	Evaluation by repeated subtraction	114
	8.4.4	Chunking	114
	8.4.5	Division by 10	115
	8.4.6	Division by 5	116

9. Regrouping — 121

9.1	Decimal units		121
9.2	Place-value notation		122
	9.2.1	Zero as a place holder	124
9.3	Regrouping		125
	9.3.1	Non-standard decimal representations	125
9.4	Numeral systems and complexity		126

10. Addition of Multi-Digit Numbers — 129

10.1	Addition without regrouping	129
10.2	Vertical addition without regrouping	130
10.3	Addition with regrouping	132
10.4	Vertical addition with regrouping	134

11. Subtraction of Multi-Digit Numbers — 139

11.1	Subtraction without regrouping		139
11.2	Vertical subtraction without regrouping		140
11.3	Vertical subtraction with regrouping		141
	11.3.1	First regroup and then evaluate	142

		11.3.2	The standard algorithm	143
		11.3.3	Multiple regroupings	143
	11.4	The French algorithm	146

12. Give Me Five!   151

- 12.1 Quinary numeral system 151
  - 12.1.1 The shepherd's tale revisited 151
  - 12.1.2 A new numeral system 152
  - 12.1.3 Place-value notation 154
- 12.2 Addition 155
  - 12.2.1 Addition of multi-digit numbers without regrouping 156
  - 12.2.2 Addition of multi-digit numbers with regrouping 157
- 12.3 Subtraction 158
- 12.4 Parity 158
- 12.5 Multiplication 159
- 12.6 Division 160

13. Introduction to Geometry   163

- 13.1 Euclidean geometry 163
- 13.2 Elementary school geometry 164
- 13.3 Set-theoretic concepts 165
- 13.4 Three-dimensional space 165
- 13.5 Geometric figures 166
- 13.6 Congruence 167
- 13.7 Measurements 168

14. Planes and Lines   171

- 14.1 Planes 171
  - 14.1.1 Plane geometry 172
- 14.2 Lines 172
- 14.3 Postulates and theorem 172
  - 14.3.1 Three Euclidean postulates 172
  - 14.3.2 A sample theorem 175
  - 14.3.3 Ordering of points on a line 176
- 14.4 Segments and rays 177
  - 14.4.1 Line segments 177
  - 14.4.2 Rays 177
  - 14.4.3 Broken lines 178
- 14.5 Segment arithmetic 179
  - 14.5.1 Segment comparison 179
  - 14.5.2 Segment addition 179
  - 14.5.3 Segment subtraction 180
  - 14.5.4 Segment multiplication 181
  - 14.5.5 Segment division 181

15. Length 183

- 15.1 Length comparison 183
  - 15.1.1 Comparison by juxtaposition 183
  - 15.1.2 Comparison by transitivity 184
  - 15.1.3 Comparison by concatenation 184
- 15.2 The length of a segment 185
  - 15.2.1 Standard measuring units 186
  - 15.2.2 The meter 186
  - 15.2.3 Using multiple measuring units 187
  - 15.2.4 Systems of measuring units 187
  - 15.2.5 Length-measuring instruments 188
- 15.3 The length of curves 189
  - 15.3.1 The length of a broken line 189
  - 15.3.2 The length of more complicated curves 190
  - 15.3.3 Infinitely long curves 191

16. Angles 193

- 16.1 What is an angle? 193
- 16.2 Angle arithmetic 195
  - 16.2.1 Angle comparison 195
  - 16.2.2 Angle addition 197
  - 16.2.3 Angle subtraction 197
  - 16.2.4 Angle multiplication 198
  - 16.2.5 Angle division 198
- 16.3 Angle measurement 200
  - 16.3.1 The degree 200
  - 16.3.2 Types of angles 201
  - 16.3.3 The protractor 201
- 16.4 Parallel lines 202

17. Polygons 207

- 17.1 What is a polygon? 207
- 17.2 Triangles 210
  - 17.2.1 The sum of the angles in a triangle 210
  - 17.2.2 Classification of triangles according to their sides 211
  - 17.2.3 Classification of triangles according to their angles 212
  - 17.2.4 Inclusion relations between types of triangles 213
  - 17.2.5 Congruent triangles and constructions 215
- 17.3 Quadrilaterals 218
  - 17.3.1 The sum of the angles in a quadrilateral 218
  - 17.3.2 Squares 219
  - 17.3.3 Rectangles 219
  - 17.3.4 Rhombuses 220
  - 17.3.5 Parallelograms 220

		17.3.6	Trapezoids	221
		17.3.7	Kites	221
	17.4	General polygons.	222
		17.4.1	The sum of the angles	222
		17.4.2	The number of diagonals	223
		17.4.3	Regular polygons	224

18. Area   227

    18.1   The area concept 227

    18.2   Area comparison 228

    18.3   Area measurement 231

    18.4   The area of polygons 232

           18.4.1   Rectangles 232

           18.4.2   Triangles 233

           18.4.3   General polygons 235

    18.5   Area and scaling 235

Index   239

Common Core Index   245

Chapter 1

Counting and Natural Numbers

The main subject of elementary school mathematics is numbers. Children's acquaintance with numbers is a long and gradual process. Their first encounter is with **natural numbers**, which are those numbers used for both counting and ordering. Over the next few years, children discover **fractions**, which are numbers that represent parts of a whole, and can be represented as quotients of natural numbers. In middle school, the world of numbers is expanded to include **negative numbers** as well. Together, whole numbers and fractions, both positive and negative, constitute a set of numbers called **rational numbers**. During their middle school years, children also get acquainted with **irrational numbers**, which are numbers that cannot be represented as quotients of whole numbers, e.g., $\sqrt{2}$. This completes the children's acquaintance with the set of numbers known as **real numbers**, which are the numbers with which we express results of quantitative measurements. In high school, teenagers pursuing advanced mathematics classes will be exposed to an even more expanded set of numbers, known as **complex numbers**. The focus of this chapter is the central role of natural numbers in quantifying the size of sets of objects.

1.1 The natural numbers

1.1.1 Development of the number concept

Counting has become such a basic skill for us that it is hard to appreciate its underlying complexities. The historic process leading to the development of the contemporary number concept is tens of thousands of years old. To the best of our knowledge, human's first use of numbers occurred about 30,000 years ago.

Long before humans invented numbers, they had to develop some manner of abstraction that would allow them to identify and name objects. Indeed, without identifiable objects there is nothing to count. Let's assume that there had already been a word for describing an object such as *a sheep*, and prehistoric men wanted to tell their friends that they had seen two sheep. They could repeat the word for sheep twice. But what would they do if they saw five sheep? They could say *sheep and sheep and sheep and sheep and sheep*, but that's awkward, and it becomes even more so if one wants to talk about a whole herd of sheep. This mundane example clarifies why, from the moment there were nouns, i.e., words that classify objects into different categories, there was a need to quantitatively describe the presence of more than one object of the same category.

1.1.2 Matching between sets

Consider the following figure, which displays two sets of objects, one kind contains *bees* and another kind, *flowers*:

Imagine that numbers have not yet been invented. What does the set containing *bee, bee, and bee* have in common with the set containing *flower, flower, and flower*? Can you say it without using the word *three*? This is not a trivial question, especially if we look for an answer that is general enough to remain valid if the bees were replaced by rabbits, and three was replaced by twenty-six.

The answer is illustrated below:

Every object in one set has been matched to a single object in the other set, without any leftover objects in either set. In mathematics, such a matching between sets is called a **bijective matching**. In this book we will use less formal language: we will either say that there exists an **exact matching** between the two sets, or that the two sets can be matched.

With this new concept in hand, we may now define what it means for two sets to be of equal size:

> **Definition 1.1.** Two sets of objects are said to be **equal in size** (or have equal size) if there exists between them an exact matching.

Please note that this way of determining whether two sets of objects are equal in size does not require counting nor assigning numerical values to the size of those sets. It also does not depend on the nature of the objects in both sets.

Let's return to prehistoric men. It is quite possible that even without being aware of the concept of exact matching, prehistoric men used to perform such a matching between sets of objects and the fingers of their hands. Thus, for example, if they said the word sheep and put up three fingers at the same time, their listeners would infer that they were describing a group of sheep that can be matched with the fingers shown. This practice would eventually pave the way for the invention of a word expressing the quantity three. The word *three* replaces putting up three fingers.

1.1.3 Counting

What do we do when we count objects? What is the meaning of the process in which we point at a collection of objects one after another and mumble: *one, two, three,* ...?

What we do, without being aware of it, is to create an exact matching between the objects in the set and a set of words—numbers—that we learned to recite as young children. These numbers form a sequence—the **number sequence**—meaning that they have a fixed order. When we match objects in a set to those numbers, the numbers have to be invoked one after the other in this order, until we have exhausted all the objects in the set.

The figure below exhibits a set of objects of the *apple* kind matched to elements from the number sequence. By the end of the matching process in which every object has been matched to a number, the last number invoked represents the size of the set (the formal mathematical term for *size* is **cardinality**).

 On natural numbers and counting

(1) **A sophisticated invention**: The numbers we use for counting are called natural numbers. German mathematician Leopold Kronecker (1823–1891) is credited for the famous saying that natural numbers are the creation of God, whereas all the other numbers are the creation of mankind.

(2) **Connecting counting to cardinality**: Counting is usually learned at a young age. Yet, children need quite some time to internalize the fact that in order to count correctly, they have to account for all the objects in the set, and that each time they point to an object, they should say one, and only one, word from the number sequence. Young children sometimes count correctly without understanding the meaning of their action, that is, without realizing that by counting they are in fact quantifying the size of a collection of objects.

(3) **An essential skill**: Counting is the first mathematical skill that children should acquire. This skill must be practiced and honed all along the early school years. One shouldn't underestimate it, for it is the basis for all future mathematical learning.

(4) **Numerals**: Numbers have names as well as symbolic representations: both the name of a number and its symbolic representation are called **numerals**. In

addition to the skill of counting, children need to make the connection between a set of objects, the numeral that expresses the size of that set, and its symbolic representation:

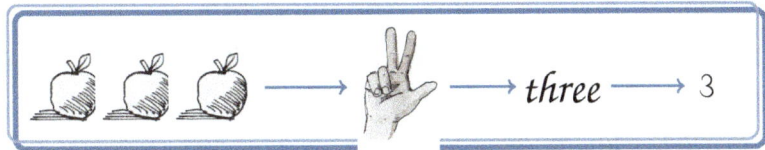

(5) **Counting at-a-glance**: I highly recommend the development of the skill of identifying, in the blink of an eye, the size of sets that include up to five or six objects. Beyond enhancing the child's confidence, which is in itself an important goal, this skill becomes useful when learning to perform mathematical operations within the range of zero to ten. This ability can be developed through games in which children are briefly shown sets of objects, and then asked to determine how many objects are in each set.

(6) **Order does not affect cardinality**: When counting objects, the result is independent of the way the objects are positioned, or of the order in which we choose to count them. This may sound obvious to an adult, but needs to be clearly pointed out to children.

(7) **Abstract versus concrete**: Numbers are abstract entities, independent of the concrete objects that they quantify. One of the central themes in mathematics is the abstraction of concrete situations. Abstraction allows us to reason about reality with better efficiency and clarity. This gap between the concrete and the abstract is one of the main hurdles a child has to overcome. For example, when children are required to find the sum of three plus two, we expect them to know that the outcome is independent of the objects to which these numbers refer. This skill is often taken for granted, but in fact, it takes a high level of intellectual maturity to treat numbers independently of the objects they quantify.

1.1.4 The number zero

Zero is a number that describes the size of an **empty set**—a set that contains no objects. The number zero has an interesting history. Western culture, influenced by Indian mathematicians, adopted zero as a number only in the 12th century CE. Children nowadays understand the meaning of zero elephants just as naturally as they understand the meaning of two elephants. Yet, it took humanity a long time to accept zero as a number representing the size of a set of objects just like any other number. For generations, people refused to assign any quantifiable trait to an empty set, arguing that *nothing* could not be counted.

Zero is not a natural number. It belongs to the set of **whole numbers**, which includes the natural numbers, $1, 2, 3, \ldots$, their negative inverses, $-1, -2, -3, \ldots$, and zero. The decision not to include zero in the set of natural numbers is purely conventional, and has no practical implication.

1.2   Order relations between numbers

1.2.1   Comparison between sets

The matching between items in two sets is a more basic concept than counting. When we are given two sets of items, we can determine whether the two sets are of equal size even if we can't count. What we have to do is to try and match each object in one set with an object in the other set. If we are successful in obtaining an exact matching, then both sets are equal in size. If one set has some leftover objects, we say that this set contains **more** objects, or that it is a **larger** set. Equivalently, we say that the other set contains **fewer** objects, or that it is a **smaller** set.

The "smaller than/equal to/larger than" relation between sets satisfies the following three properties:

(1) Every set is equal in size to itself.
(2) Every two sets, which we name Set A and Set B, satisfy one, and only one, of the following three possibilities:
   (a) Both sets are equal in size.
   (b) Set A is smaller than Set B (equivalently, Set B is larger than Set A).
   (c) Set B is smaller than Set A (equivalently, Set A is larger than Set B).
(3) If Set A is smaller than Set B, and Set B is smaller than Set C, then Set A is necessarily smaller than Set C. This property is called **transitivity**, which means, literally, passing on a property from one instance to another.

In mathematical jargon, a relation between mathematical objects that satisfies the above properties is called an **order relation**. In the present context, those mathematical objects are sets. Sets satisfy an order relation. Please note that the order relation between sets does not depend on the identity of their elements. Every item has the same level of importance, regardless of its type, its physical size or any other feature.

 **Activity: Counting and comparing**

(1) Present children with two sets of objects. Ask them which set is larger. It is very likely that they will start counting the objects in each set.

(2) Stop them from counting, and ask them to answer the question without counting. For example, let them pretend that they are prehistoric humans, who live in the era before numbers were invented.

1.2.2 Comparison between numbers

The order relation among sets induces an order relation among natural numbers. We say that Number A is **greater** than Number B if a set that contains A objects is larger than a set that contains B objects. In other words, the assertion that *five is greater than three*, which is an assertion about numbers, means: if we try to match items from a set of five objects with items from a set of three objects, unmatched items will remain in the set of five.

The order relation between numbers satisfies the properties satisfied by all order relations. Specifically:

(1) Every number is equal to itself.
(2) Every two numbers, which we name A and B, satisfy one and only one of the following three possibilities:
   (a) Both numbers are equal.
   (b) A is smaller than B (equivalently, B is greater than A).
   (c) B is smaller than A (equivalently, A is greater than B).

(3) If A is smaller than B, and B is smaller than C, then A is necessarily smaller than C.

Mathematics always prefers succinct notations to words. The greater/smaller relation between numbers is expressed using **mathematical sentences**. The mathematical symbol for the relation **greater than** is >. The mathematical symbol for the relation **less than** is <. The signs > and < are called **relation signs**.

For example, the mathematical sentence that expresses the assertion that *8 is greater than 5* is

$$8 > 5,$$

whereas the mathematical way to express the assertion that *5 is less than 8* is

$$5 < 8.$$

1.2.3 Successor and predecessor

A distinctive feature of the natural numbers is that each number has a **successor**, which is the natural number that succeeds it in the number sequence. The successor of a number represents a set that has one more object. Formally speaking, the successor of a natural number is the smallest natural number among all those numbers that are greater than it is. For example, 6 is the successor of 5, because it is the smallest natural number among all those natural numbers that are greater than 5.

Similarly, the **predecessor** of a natural number is the natural number that precedes it in the number sequence. The predecessor of a number represents a set that has one less object. Formally speaking, the predecessor of a natural number is the greatest natural number among all those numbers that are less than it is. For example, 4 is the predecessor of 5, because it is the greatest natural number among all those that are less than 5.

Every natural number has a successor, but there exists a distinctive natural number that has no predecessor—1. The number 1 is the smallest of all natural numbers (recall that zero is not considered a natural number). The existence of successors and predecessors is a fundamental characteristic of natural numbers. The set of rational numbers (i.e., the set of numbers that includes fractions) no longer comprises a notion of successor and predecessor. For example, there exists no number that is the smallest rational number among all those that are greater than five.

1.2.4 There are infinitely many natural numbers

Even preschool children realize that the number sequence has no end. This means that there are infinitely many natural numbers. What we mean by that is if we were to list the natural numbers in ascending order, this process would never end; every newly added natural number would have a successor that had not yet been listed.

A common misconception among children is that there is a number called **infinity**, greater than all other numbers. Infinity is not a name of a number.

1.2.5 The number line diagram

A convenient graphical tool for expressing the order relation between numbers is to represent them on a straight line. Such representation is called a **number line diagram**. It has many didactic uses, and we will encounter some of them along the book. As for now, we only explain the principles underlying this representation.

To place numbers on a line, we first choose an arbitrary point on the line and identify it with the number 0. We next choose an arbitrary point to its right and identify it with the number 1. That ends the arbitrariness in the choice of points. We then mark, to the right of 1, another point at the same distance from 1 as the distance between 0 and 1. We identify it with the number 2. We keep marking points on the line, each new points at the right of its predecessor, fixing the distance between successive points. We finally assign to each point a natural number, in ascending order from left to right:

The order relation between numbers has a simple expression on the number line: a number is greater than another if it is located to its right. Please note that the number line diagram emphasizes the order relation between numbers. At the same time, it obscures the role of numbers as representing the size of sets.

1.3 Natural numbers as determining order

The main role of the natural numbers is to quantify sizes of sets. The natural numbers have another important role: to determine the order of the items within a set.

Every collection of objects can be assigned an order. When we count objects in a set, we assign each object a natural number. If we are only interested in knowing how many objects there are, then the assignment of numbers to objects is only an auxiliary tool; we do not care which number was assigned to which object. If, however, we maintained the assignment of numbers to objects, the order between the numbers would create an order between the objects.

Creating an order between objects is part of our daily routine. At a very young age, children learn to exclaim: *I'm first*! What does it mean to be first? If we matched a group of children with a sequence of natural numbers, the *first* child would be the one matched with the number 1. Likewise, people instructed to *take a right at the third light* understand the instruction very well. They will turn at the third light after having counted the lights on their way, assigning the number 1 to the first light, assigning the number 2 to the next, and assigning the number 3 to the next. Once again, we see the fundamental importance of matching.

1.4 Estimation

We often want to know how many objects are in a set, but rather than counting the objects we are satisfied with an approximation of the actual quantity. Such an approximation is called an **estimate**. The making of an estimate is called **estimation**.

There are several reasons that we sometimes settle for an estimate rather than an exact count. Sometimes it is just difficult to count. If you were asked how many hairs you have on your head you would obviously find it very difficult to count them (unless you are bald). Sometimes we settle for an estimate because we are not interested in the precise number. For example, we would probably be content to know that about one and a half billion people live in China, rather than knowing that their exact number is 1,420,546,792 (not to mention the fact that this number is changing as we read this sentence).

The notion of "good" or "poor" estimate is not absolute. The quality of an estimate is context dependent. Returning to the population of China, while the reader of this book will settle for an estimate of one and a half billion, the Chinese Minister of the Interior needs more exact figures.

The ability to estimate is very important in everyday life. A person who is planning to travel has to estimate the cost of the trip without knowing every single expense in advance; someone who is planning to paint a house should be able to estimate the amount of paint needed before setting to work. The ability to make good estimates develops gradually and requires practice.

 **Activity: Estimation versus counting**

(1) Present children with an assortment of objects, such as a box of beads.

(2) Ask each child to estimate the number of objects. This can be presented as a guessing game.

(3) After the estimation process has been completed and all the estimates have been recorded, ask the children to count the objects, thus revealing their exact number and connecting the process of estimation to the process of counting.

(4) This activity can be repeated with objects of different shapes and sizes, to show how shape and size may affect estimation but not counting.

1.5 More on numbers and sets

1.5.1 What is *three*?

Mathematics builds up theories from elementary concepts, which are taken for granted without further definition. A set, i.e., a collection of objects, is an example of such an elementary concept. In **set theory** (which is the area of mathematics studying sets), two sets are considered **equivalent** if their elements can be matched exactly (as the bees and the flowers on p. 2).

Equivalence has a feature we called **transitivity**. If Set A is equivalent to Set B, and Set B is equivalent to Set C, then Sets A and C are also equivalent. The concept of equivalence divides all sets into **equivalence classes**, that is: classes of sets that are all equivalent to each other.

For example, the set
$$\{\text{Summer, Winter}\}$$
belongs to the same equivalence class as the sets
$$\{\text{fish, owl}\}, \quad \{\text{broom, mop}\} \quad \text{and} \quad \{\text{David, Goliath}\},$$
because the elements of any two sets in this collection can be matched exactly. But the set
$$\{\text{James, Isabel, Juan}\}$$
belongs to another equivalence class, since it cannot be matched exactly to any of the other sets. In mathematics, *three* is the name of the equivalence class that includes the set {James, Isabel, Juan}, as well as the set {earth, wind, fire}.

1.5.2 Comparison of infinite sets

The concept that two sets are equivalent if they can be matched exactly is also used to compare sets that contain an infinite number of elements. Consider the following two sets:

$$\text{Set A} = \{\text{all the multiples of 2}\},$$
$$\text{Set B} = \{\text{all the multiples of 3}\}.$$

Both sets have an infinite number of elements. We can nevertheless find an exact matching between these two sets. To see how, let's examine the figure below:

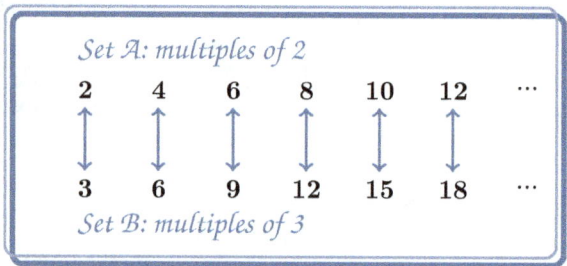

Each element in Set A can be matched with one, and only one, element in Set B, such that there is no element left unmatched in either set. For example, 2 in Set A is matched to 3 in Set B, 4 in Set A is matched to 6 in Set B, and so on. By our definition, the two sets are equivalent. Even though Sets A and B can be matched exactly, we do not call them sets of equal size—the term *size* is reserved for finite sets. When two infinite sets can be matched, the mathematical terminology is that they both have the same **cardinality**.

The concept of cardinality of infinite sets, developed in the second half of the 19th century, expands the concept of the size of finite sets. One of the surprising implications of the definition of cardinality is that an infinite set can have the same cardinality as one of its own subsets. For example, the set of even numbers is a subset of the set of all natural numbers. We would be inclined to say that there are twice as many natural numbers as even numbers. Yet, the set of even numbers has the same cardinality as the set of natural numbers, as every natural number can be matched with the even number that is twice as large. Infinite sets, then, do not comply with one of the premises dating back to ancient Greece: that the whole is always larger than any one of its parts. Another surprising fact is that not all infinite sets have the same cardinality; different magnitudes of infinity exist.

Mathematical problems

Problem 1.1. Can you assert with certainty that there exist at every moment two people in Chicago who have the *exact* same number of hairs on their head? (You may exclude bald-headed people.)

Problem 1.2. Do the following two sets have the same cardinality?
$$\text{Set A} = \{1, 2, 3, \dots\},$$
$$\text{Set B} = \{\dots, -2, -1, 0, 1, 2, \dots\}.$$
Justify your answer based on the notion of exact matching.

Chapter 2

The Decimal Representation System

Imagine that we were prehistoric humans, inventing natural numbers. Every time we wanted to express a number for which a numeral, i.e., a name, had not yet been invented, we would have had to come up with a new numeral. To be able to express an infinite sequence of natural numbers would require an infinite vocabulary. One of the greatest human inventions ever is the ability to express any natural number with a finite collection of symbols, which we call **digits**.

The difficulty of generating more and more numerals to represent an expanding range of numbers must have emerged once the use of numbers became a human routine. The first known numeral system to confront this difficulty was invented by the Babylonians around the year 4000 BCE. Since then, many other numeral systems have been invented in different parts of the world. The system we now use is called the **decimal system**. In this chapter, we will get acquainted with its underlying principles. A deeper study of the decimal system is presented in Chapter 9.

2.1 Development of the decimal system

2.1.1 The legend of the shepherd

It can be assumed that at some point in history humans learned to count to ten, since this number can be represented with the fingers of both hands. In order to understand how they coped with the need to name and represent more numbers, consider the following hypothetical, yet very likely scenario:

A prehistoric shepherd wants to count the sheep in his herd. He does not know how to do that, since the herd is big and he can only count to ten. What can he do? He takes a jar, and every time he comes across a sheep, he drops a pebble into the jar. When he returns home, he pours the pebbles out of the heavy jar, knowing that each pebble represents a sheep in his herd. (Question: Can the shepherd tell another person how many sheep there were without using the pebbles?)

The next time the shepherd wants to count his herd, he has a better idea: In order to avoid carrying a heavy jar home again, he decides that every time he would count ten sheep—which he can do—he would drop a *single* pebble into the jar. He knows that each pebble does not represent one sheep, as the previous time, but ten sheep. However, when he is almost done counting, he realizes that there are three sheep left that he has not yet counted. He considers adding three pebbles to the jar, but rejects the idea, since a pebble represents now ten sheep.

What can he do? After much thought, the shepherd decides to add three twigs to his jar, so that it would be easy for him to distinguish between twigs which represent one sheep each and pebbles which represent ten sheep each. When he comes home he puts down the jar, which is significantly lighter this time, and sees that it contains seven pebbles and three twigs.

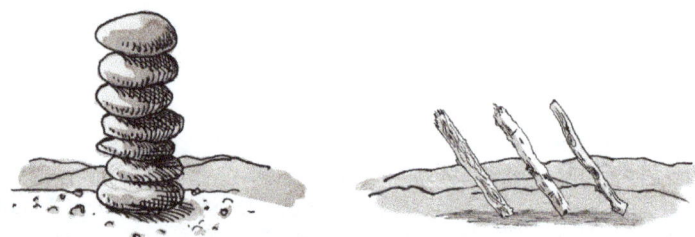

2.1.2 Ancient decimal representation

In this section, we describe how the shepherd from the tale ended up developing a numeral system to represent the natural numbers. The description is imaginary, and yet, very similar to the actual way numeral systems developed in various cultures (see Section 2.3 ahead).

In the beginning of the tale, the shepherd used one type of objects—pebbles—to represent set sizes; one pebble for each object in the set. His first numeral system imitated this practice, replacing the material pebbles by symbols. For example, suppose he used vertical segments, one vertical segment for each object in the set. He represented the number sequence using the following numerals:

$$| \quad || \quad ||| \quad |||| \quad ||||| \quad \ldots$$

Every | represents one object. A numeral system based on an exact matching between a collection of objects and a collection of symbols is called a **unary numeral system** (because it only uses one digit).

Later on, once the shepherd developed a more efficient counting system, matching a pebble with ten objects and a twig with single objects, he needed two sorts of symbols: one for twigs and one for pebbles. The shepherd continued to use the symbol | to represent one object, but he added a new symbol, •, which has the same value as ten | (i.e., a | replaces a twig and a • replaces a pebble). Today we call the quantity which is represented by the symbol • a **ten**. The action of converting ten identical entities by a single entity whose value is ten times larger is called **bundling**, or **grouping**. In this case, ten ones are bundled to form one ten.

The numerals following • are represented by a combination of • and | (pebbles and twigs):

$$•| \quad •|| \quad •||| \quad •|||| \quad \ldots \quad •|||||||||$$

The numeral •||||||||| represents a combination of a ten and nine ones (one pebble and nine twigs). To represent its successor the shepherd would have to add another |. Then he would get ten |, which he would bundle into a •, thus obtaining two •. The number sequence proceeds as follows:

$$••| \quad ••|| \quad ••||| \quad ••|||| \quad \ldots \quad ••||||||||| \quad ••• \quad •••| \quad \ldots$$

Had the shepherd continued writing numerals in increasing order, he would have reached eventually the following numeral:

$$•••••••••|||||||||$$

which represents nine tens and nine ones (nine pebbles and nine twigs). In order to represent the next number he would have to add a |. Once again he would have ten | which he would bundle into a •. With ten •, the resourceful shepherd decided to apply the principle once again, whereby every ten identical items are bundled into a new entity whose value is ten times larger. He therefore introduced a new symbol ▲ that stands for the same quantity as ten •. Today we call the number represented by the symbol ▲ a **hundred**. Thus, he bundled ten tens into one hundred.

For example, in the shepherd's numeral system, the sequence of symbols

$$▲▲••••|||||$$

represents the size of a set containing five objects, and four tens of objects, and two hundreds of objects, where one ten equals ten ones, and one hundred equals ten tens. Today, the counting units *one*, *ten* and *hundred* are called **decimal units**.

2.1.3 Place-value notation

The main constituent of the decimal numeral system is the bundling of ten identical decimal units into a decimal unit whose values is ten times larger. The first such system was introduced in ancient Egypt, about 5000 years ago (see also p. 22). Several other numeral systems based on bundling tens have been developed throughout history, differing in the way the numbers were represented. The numeral system that was eventually adopted is distinctive in its notational compactness. The first step in its development was the introduction of ten symbols—**digits**—representing the whole numbers between zero and nine. The digits used in most of the world today are called the **Hindu-Arabic digits**,

$$0, 1, 2, 3, 4, 5, 6, 7, 8, 9.$$

With ten digits at our disposal, we could combine the digits with symbols representing the decimal units, such as |, • and ▲ in the shepherd's tale. Instead of •••••••• for the number *eight tens*, we could write 8•, and instead of ▲▲••••||||| for the number *two hundreds, four tens and 5 ones*, we could write 2▲ 4• 5|.

Now comes the climax of the decimal representation system: the omission of the symbols |, • and ▲. We only write a sequence of digits, and conform to the following convention: the digit on the far right counts ones, the one to its left counts tens, and the one to its left counts hundreds. For example, instead of ▲▲••••|||||, we write 245. We call this convention the **place-value principle**, since the value of a digit depends on its place within the numeral.

2.1.4 Zero as a place holder

Consider, for example, a number that comprises 5 hundreds and 7 ones. In the fictional representation system that preceded the use of the place-value notation, we would have written it this way:

$$5▲ \ 7|$$

If we were to omit the decimal units ▲ and |, we would get the sequence of digits 57. However, according to the place-value convention, 57 represents 7 ones and 5 tens (instead of 7 ones and 5 hundreds).

Although the decimal representation of the number that comprises 5 hundreds and 7 ones does not include tens, the place-value convention requires the use of a *place holding digit* in the place designated for the tens digits. The natural digit to choose for this task is 0. Thus, it is the sequence of digits 507 that represents 7 ones, (no tens), and 5 hundreds.

Benefits of the decimal representation

The decimal numeral system has three major virtues:

(1) **Succinctness**: Look back at the shepherd tale: due to his improved counting system that combines pebbles and twigs, the jar is much lighter. The lighter weight of the jar is a metaphor for the more efficient writing that characterizes the decimal representation. In the decimal system, the representation of a number that is ten times larger than a given number only involves the addition of one digit. Compare the system of decimal representation with a unary numeral system in which each individual object is represented by a symbol. You will easily see how futile it is to attempt writing the number *one million* without an efficient numeral system.

(2) **Structure**: The decimal numeral system offers much more than just an efficient representation of large numbers. When we bundle ones into tens and tens into hundreds, we create a structure. Consider the following figure:

On the left are 14 scattered matches. On the right the same 14 matches are arranged in a structure that imitates the principles of the decimal system: Ten matches are bundled together to form a new object that is a ten-of-matches. That bundled ten-of-matches is placed to the left of the four remaining matches, which remain scattered. The figure displays (from left to right) 1 ten and 4 ones, just as in the numeral 14.

(3) **Quantitative insight**: The structure imposed by the decimal representation is essential to our ability to grasp large quantities and perform mathematical operations on them. Think for a moment of a number such as 328. What is it we imagine? 328 scattered objects? Obviously, we would not be able to distinguish between 328 scattered objects and, say, 335 scattered objects. Such quantities must be organized in a structured manner for us to be able to distinguish between close numbers, and develop quantitative insight.

2.2 Early age learning of the decimal system

When young children recite the counting sequence, *one, two, three, ..., nine, ten, eleven,...* they do not feel that anything has happened in the transition from nine to ten. Children are exposed to the decimal representation system long before they reach school age, but usually without being aware of it. Many children can count to a hundred and beyond, even before first grade, and some know how to read multi-digit numbers without understanding the principles underlying this numeral system.

The American Common Core Standard introduces the decimal representation gradually. In fourth grade, children learn the principles underlying the representation of multi-digit numbers. Long before that, they learn to evaluate arithmetic operations with 2-digit, 3-digit and 4-digit numbers. In light of this, one of the teaching goals in the lower grades is to facilitate by example the internalization of the numeral system in children.

This can be done in a number of different ways, including the following two strategies:

(1) Every time you manipulate more than nine objects, organize them (or ask the children to organize them) in a way that conforms to the decimal representation, as illustrated in the figure on p. 19. Translating the abstract representation into a concrete structure of objects creates a mental connection between representation and interpretation.

(2) Use every opportunity to refer to the meaning of the decimal representation. For example, if the result of a certain calculation is 37, one can say: *the result is thirty-seven, which is three tens and seven ones. Therefore, we write 3, which represents the number of tens, and to its right 7, which represents the number of ones.* Using such sentences routinely contributes to the internalization of the decimal representation. In general, weaving explicit explanations into the flow of one's teaching is a good way to inculcate an idea in a short amount of time without boring children.

 **Activity: Structured counting**

Present children with a pile of hundreds of sticks and ask them to count them in a way that conforms to the decimal representation. Have them:

(1) Bundle together every ten sticks until they are left with less than ten scattered sticks.
(2) Bundle together every ten bundled tens until they are left with less than ten scattered tens.

(3) Display the sticks in the following way: the scattered sticks on the right, the bundled tens to their left, and the bundled hundreds to the left of those.
(4) Summarize the count explicitly: e.g., we have 4 hundreds, 3 tens, and 8 ones.
(5) Write down the digits representing the decimal units from left to right (4, 3 and 8 in the above example).

2.2.1 The hundred table

The hundred table is a teaching aid for getting acquainted with numbers. This table features all the natural numbers between zero and one hundred, and they are arranged in a square grid, as shown below:

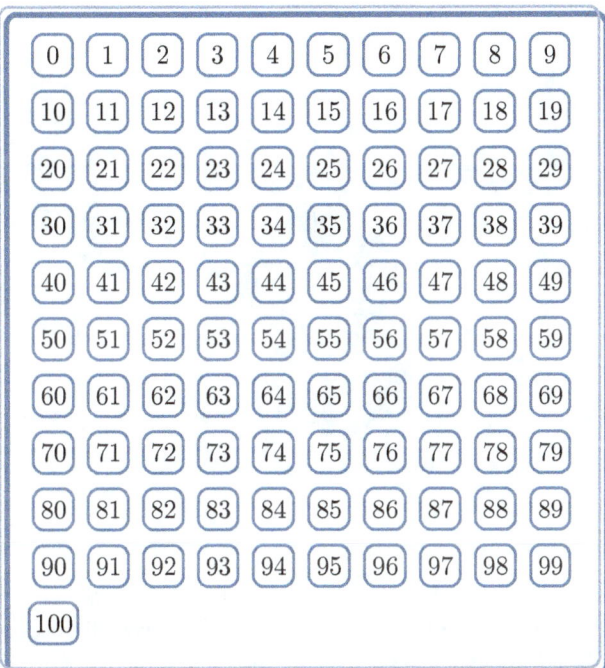

Recall the number line diagram that illustrates the order relation between numbers (p. 9). Here, the numbers are organized differently. Numbers in the same row have the same tens digit. Numbers in the same column have the same ones digit. A move to the right, to the left, up or down corresponds to the addition or subtraction of a one, or a ten. The hundred table provides a visual perspective on a relatively large set of numerals.

The hundred table offers many opportunities for playing number games to help children connect between the magnitude of a number and its numeral representation. For example:

(a) Skip count in intervals of 2, and color the numbers in this sequence. What pattern do you observe? Explain.
(b) Repeat the same with intervals of 10, 5, 9 and 11.
(c) Color all the numbers whose tens digit is larger than the ones digit. What pattern do you observe?
(d) Play guessing games: one child chooses a number and another child has to find that number with a minimal number of yes/no questions.
(e) Provide children with a partially filled hundred table and let them complete it.

2.3 Numerals in other cultures

This section describes some of the numeral systems developed in different cultures: ancient Egypt, China, and Rome.

2.3.1 Ancient Egypt

There is evidence from as early as the year 3100 BCE to the existence of a decimal numeral system in ancient Egypt. The Egyptians used seven digits to represent decimal units from one to a million, as shown in the following table:

In order to represent a number, the Egyptians would repeat each decimal unit as many times as necessary. For example, the number 43051 was written this way:

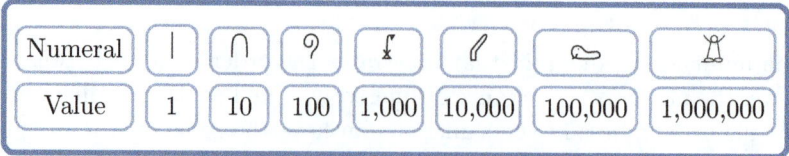

The Decimal Representation System

Note that this numeral system is a decimal system that does not use a place-value notation; every digit has a fixed value regardless of its position. The representation of a number in this system would almost always necessitate more digits than in our current system. For example, in the Egyptian numeral system the representation of the number 9999 requires 36 digits.

2.3.2 China

The Chinese numeral system is also decimal. Like the Western system, there are nine non-zero digits, from 1 to 9:

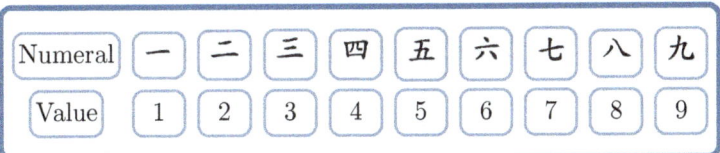

As in ancient Egypt, there are also digits that represent the decimal units:

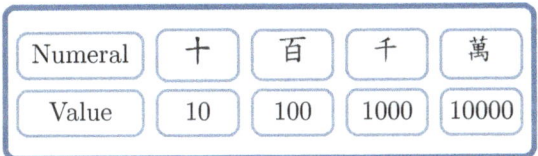

The number 25074, for example, is written this way (from left to right):

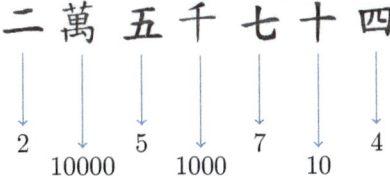

Please note that in this system, there is no need for a place holder digit. Also, the value of a digit between 1 and 9 does not depend on its position, but rather on the identity of the decimal unit located to its right.

2.3.3 Ancient Rome

A decimal numeral system was also prevalent in ancient Rome. The numerals were represented by letters, as in the table below:

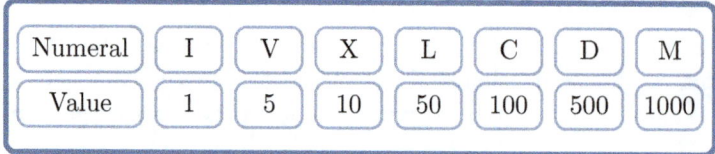

When digits are arranged in descending order from the largest to the smallest, the value of the numeral is the sum of the digits that make it up. For example:

$$302 = \text{CCCII} \qquad 32 = \text{XXXII}.$$

When a digit of lower value is placed to the left of a digit of higher value, the digit of lower value is subtracted, rather then added. For example:

$$744 = \text{DCCXLIV} \qquad 9 = \text{IX}.$$

Mathematical problems

Problem 2.1. Write the numbers 507 and 1444 using the Egyptian, Chinese, and Roman numeral systems.

Chapter 3

The Four Operations of Arithmetic

A major part of elementary school mathematics is dedicated to the four fundamental operations of arithmetic—**addition**, **subtraction**, **multiplication** and **division**. This chapter provides a general introduction to all four operations. Each operation is studied in detail in later chapters.

3.1 Arithmetic operations

Like all mathematical concepts, arithmetic operations can be defined in very formal terms. However, we use a figurative way to describe what arithmetic operations are:

> An arithmetic operation is a machine with two **input** ports and one **output** port. When two numbers are entered into the machine's input ports, the machine computes a number and returns it as its output.

The figure below presents an **addition machine**. The two input ports, marked A and B, are to the left of the machine. The output port is on its right. When we insert the numbers 5 and 6 into the machine, the result of the operation, 11, appears in the output port.

A given machine always returns the same output if provided with the same input. What distinguishes between the different arithmetic operations is the rule by which the machine generates the output, given the input. If we inserted the same numbers, 5 and 6, into a multiplication machine, we would get a different output than the one we obtained above.

3.1.1 Operators and operands

In mathematical jargon, such a machine is called an **operator**. The two input numbers are called **operands** (the objects that the operator acts upon). The operands of the four operations of arithmetic have names that are specific to each operation:

(1) **Addition**: the two operands are called **addends**. Some authors, however, distinguish between the two operands, calling the first the **augend** and the second the **addend**.

(2) **Subtraction**: the first operand is called the **minuend** and the second the **subtrahend**.

(3) **Multiplication**: the first operand is called the **multiplier** and the second the **multiplicand**. Sometimes, however, both operands are given the same name—**factors**.

(4) **Division**: the first operand is called the **dividend** and the second the **divisor**.

In the early grades, the numbers that are inserted into a machine and the result of the operation are always whole numbers. Starting from third grade, this range of values is gradually expanded to include fractions as well.

3.1.2 Concatenating operations

Suppose that we have two machines: an addition machine and a multiplication machine. Each machine has two input ports and one output port. It is possible to combine these two machines into a single compound machine that has three input ports and one output port. To do so, we **chain**, or **concatenate** the two machines: We direct the output port of one of the machines into an input port of the other machine, as in the figure below:

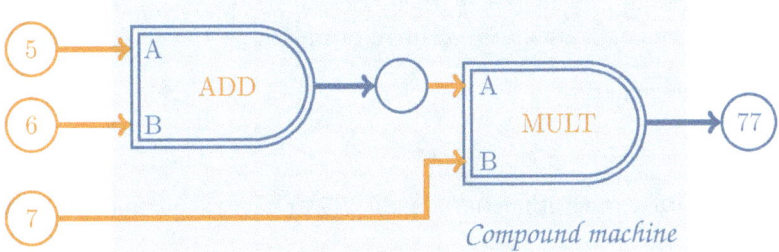

Compound machine

For every *three* numbers that we insert into this compound machine, we get a number as output. For example, when we insert the numbers 5, 6, and 7, in the order shown in the above figure, we obtain the result 77.

The concatenation of multiple operations is used extensively in the solution of multi-step problems. Moreover, some of the properties of the operations of arithmetic (e.g., the associative and the distributive properties) concern the concatenation of operations (Chapters 4, 5, 7 and 8). Multiple operations are also discussed in Volume 2, in the context of the order of operations.

3.2 Arithmetic expressions

Notation is a key theme in mathematics. Two of the distinctive traits of mathematics are abstraction and economy, and both come into play in the way we write mathematical expressions. On the down side, learning mathematical notation is similar to learning a new language: it takes time and practice. One should not underestimate the difficulty that mathematical notation constitutes for the novice.

In mathematical notation, the four machines that represent the four operations of arithmetic are denoted by the symbols +, −, × and ÷. Instead of being inserted through two input ports, the two operands are placed on both sides of the symbol that represents the operation; the left-hand side of the symbol + is Port A and the right-hand side is Port B.

For example:

$$5 + 6$$

represents an addition machine into which the numbers 5 and 6 (in that order) have been inserted. The expression $5 + 6$ is an instance of an **arithmetic expression**: a combination of numbers and operation signs that has a numerical value. In this case, the numerical value of the expression is the result produced by the addition machine when the numbers 5 and 6 are inserted into it.

To make the implicit value of an arithmetic expression explicit, we state that the arithmetic expression equals a number. To state that two expressions have the same value we use the **equal sign** =, which indicates that the two expressions written on its two sides are equal in value. A mathematical statement asserting that there is an equality between two expressions is called an **equation**.

When we write the equation

$$5 + 6 = 11,$$

we assert, in fact, that the arithmetic expression $5 + 6$ and the number 11 have the same value.

Please note:

(a) Mathematics is written from left to right, which creates difficulties to those whose native language is written from right to left (e.g., Hebrew and Arabic).

(b) Children often mistakenly think that the equal sign is an action sign (perhaps because when using a calculator, pressing = is followed by the performance of the calculation). Thus, many children tend to think that the equation

$$11 = 5 + 6$$

is written backwards. It is important to point out that this equation is correct, and is in fact identical to the equation $5 + 6 = 11$. Both equations state that $5 + 6$ and 11 have the same numerical value.

An arithmetic **evaluation problem** is a task whose aim is to **evaluate** (i.e., to find the numerical value of) an arithmetic expression. In this book, arithmetic problems are presented as equations with an unknown:

$$5 + 6 = \boxed{?}$$

The framed question mark indicates that we want to substitute it with a number such that both sides of the equation have the same value.

3.3 Teaching the operations of arithmetic

When teaching the four operations of arithmetic several points should be emphasized:

(1) **Word problems**: Children need to develop the ability to solve word problems and make the connection between concrete, real life situations, and the abstract arithmetic operations that model them. Children should know when each operation should be used. The conversion of a concrete problem into an abstract mathematical formulation is called **modeling**.

(2) **Interpretation**: Each operation has several interpretations, each with its own nuances. Understanding the interpretations of an operation is essential in order to know when to use each operation, for example, in solving word problems.

(3) **Properties**: Each operation satisfies its own properties (e.g., addition satisfies the commutative property). Recognizing and understanding those properties is an indispensable part of learning an operation. In fact, these properties become almost self-explanatory, and, therefore, easier to internalize, when one knows how to interpret an operation.

(4) **Relationships**: The four operations of arithmetic are closely related to each other. Subtraction is the inverse operation of addition, multiplication is repeated addition, and division is the inverse operation of multiplication. The relationships between operations are direct consequences of their interpretations.

(5) **Evaluation**: Eventually, children learn how to find the output of an operation, given its input (i.e., they learn to evaluate arithmetic expressions). Parallel to the development of computational skills, it is important for children to understand the principles underlying every method of evaluation.

Each of these points will be thoroughly expanded in the following chapters within the specific contexts of each operation.

Chapter 4

Addition

Addition is the first arithmetic operation learned in school. It is the most basic of all operations, in the sense that it is the most intimately related to counting. In this chapter, we get acquainted with addition, along with its interpretations and its properties. We address the evaluation of addition within the number range that first graders are expected to have mastered. The evaluation of addition for arbitrarily large operands is addressed in Chapter 10.

4.1 What is addition?

4.1.1 Addition as a model for joining together

There are two sets of objects: a set of pencils and a set of erasers. We know the number of objects in each set—there are 4 pencils and 5 erasers. We then **join together** (or **put together**, or **combine**, or **unify**) the two sets of writing utensils into one compound set. Can we tell how many objects there are in the compound set, or in the present context, how many writing utensils there are in total?

The most important aspect in questions of this kind, is that the answer does not depend on the kind of objects involved. It doesn't matter at all whether we are dealing with pencils and erasers or with hens and rabbits. Furthermore, the answer neither depends on how the objects were arranged before the two sets were joined together, nor it depends on *how* the two sets were joined together. The answer only depends on the size of the two sets, because counting does not depend on the nature of the objects counted or on the way we choose to enumerate them.

Thus, it makes sense to ask *"how much is 4 plus 5?"* without any reference to a concrete context. The solution to this problem is the number of items in a set obtained by joining together a set of 4 items with a set of 5 items. The answer, in this case, is 9.

4.1.2 Addition as a model for appending

In the previous section, we considered the joining together of two sets of objects. In this setting, the two sets have the same status: we do not distinguish between a *first* set and a *second* set. Sometimes, however, sets are combined in an asymmetric way.

Consider the following question:

> 8 children were playing at the playground. 2 more children joined them. How many children are now playing at the playground?

In this question, a quantity (2 children) is **appended**, or **added to** another quantity (8 children). The two sets do not have the same status; the situation described in this question differs from a situation where 2 children who were playing at the playground were joined by 8 children. The fact that in both cases there are eventually 10 children at the playground proves, once again, that the result of joining together two sets is independent of how the two sets were joined.

4.1.3 The addition operation

Addition is an operation that takes two numbers, which we may interpret as representing the size of two sets, and returns a number, which is the size of the set obtained by joining the two sets together. The two **operands** of the addition operation are called **addends**. Some texts choose to name the operands in a manner that emphasizes their different roles in the *adding to* setting. The first operand is then called the **augend**—the number being augmented—whereas the second operand is called the **addend**—the number being added. The outcome of the addition operation is called the **sum**.

The addition sign is called **plus**, and it is denoted by +. For example, the arithmetic expression

$$5 + 4$$

expresses the outcome of adding 5 and 4. The numerical value of the expression $5 + 4$ is the number of items in a set which is the result of joining together a set of 5 items and a set of 4 items.

The equation

$$4 + 5 = 9$$

express the equality between the arithmetic expression $4 + 5$ and the number 9.

About addition

(1) **Addition is based on counting**: Before teaching any new subject, it is important to know what prior knowledge is required to comprehend it. Addition is based on counting: it assumes that we can count items in two distinct sets, and that we can count the items in the set obtained by joining the two sets together.

(2) **Independence of specifics**: The size of a set obtained by joining together two sets of given sizes does not depend on the type of items that those sets contain. It is a fact that we have internalized from long and cumulative experience, and it should not be taken lightly. Here is a way to convey this understanding to children. Ask them: "*How many beans do you get by joining together 2 beans and 3 beans?*". Once they have answered, ask them: "*How many pencils do you get by joining together 2 pencils and 3 pencils?*". Keeps asking similar questions changing only the type of objects used. Eventually, they will realize that *2 plus 3 always equals 5.*

(3) **Apples and oranges**: The idiom *apples and oranges* is a metaphor for items that cannot be jointly quantified. This idiom has profound consequences in mathematics: quantities can be counted together, compared, added, and subtracted only if they can be considered of the same type. The question "*what do you get by joining together 3 apples and 2 oranges?*" is meaningful only because apples and oranges can be considered to be of a joint type—both are fruits. Similarly, we can add pencils and beans together only if we can endow them with the same denomination, for example, "countable objects that are currently on the table". On the other hand, it doesn't make much sense to join together 3 handshakes and 2 horses.

(4) **Joining together is not necessarily physical**: Consider the following word problem: *The third grade in the school is made up of two classes. 3A has 21 students, and 3B has 24 students. How many students are in the third grade?* The third grade in this question is the union of two classes, but in order to answer the question there is no need to actually merge the two classes.

(5) **Addition of ordinal numbers**: Addition has a different interpretation in the context of ordinal numbers. Let's look at the following problem: *Amy is fourth in line, and Jimmy is standing three spots behind her. What is Jimmy's place in the line?* In order to answer the question, we have to calculate 4 plus 3, but the two numbers in this expression have different roles. The 4 is an ordinal number (Amy's place in the line). The 3 represents the distance, or difference, between two ordinal numbers (between Amy's place and Jimmy's place). The result (Jimmy's place) is once again an ordinal number.

4.2 Word problems

Solving word problems is the raison d'être of mathematics, especially when it comes to elementary school mathematics. Numbers, operations and evaluation algorithms are abstractions developed to systematize our approach to word problems, and they take advantage of recurring patterns. Mathematics, striving to abstract away the unnecessary details, converts word problems into abstract arithmetic problems.

4.2.1 Modeling with addition

In the context of addition, modeling means identifying those situations in which the solution to the problem is the sum of two or more addends. We have already seen several such situations. For example:

(a) There are 9 pennies in my right pocket. There are 2 pennies in my left pocket. How many pennies do I have in total?
(b) I had 9 raisins. I was given 2 more raisins. How many raisins do I have now?
(c) My brother is 9 years old. I am 2 years older. How old am I?
(d) I live on the 9th floor. My friend lives 2 floors above me. What floor is she on?
(e) I am 9th in the line. My friend is two spots behind me. What place is she in line?

In all those problems, the solution is the value of the arithmetic expression:
$$9 + 2.$$

4.2.2 Interpretation

In the context of addition, interpretation means assigning a concrete situation to an abstract addition problem.

Suppose we have to solve the following arithmetic problem:
$$764 + 2 = \boxed{?}$$
To evaluate an arithmetic expression, it is often insightful to visualize a concrete situation in which such expressions may arise. The essential point is that we have the freedom to choose any appropriate interpretation. This freedom can be a real asset. For example, when we evaluate $764 + 2$, it is usually easier to imagine a situation in which 2 objects were appended to 764 objects, than a situation in which 764 objects were appended to 2 objects, or a situation in which two sets were joined together symmetrically.

4.3 The properties of addition

Addition, like all the other arithmetic operations, satisfies distinctive properties. In the following pages, we will present four properties that are intrinsic to addition: the commutative property, the associative property, the existence of an identity element, and the laws of variation. Understanding these properties is important both for their own sake—part of understanding what addition is—and for developing and applying evaluation strategies.

4.3.1 The commutative property

In mathematics, an operation that operates on two numbers is called **commutative** if changing the order of the operands does not change the result. Addition is commutative, which means that the sum does not change when we interchange the two addends. The commutative property of addition derives from the fact that when two sets are joined together, it does not matter which set comes first (see p. 32).

For example, it follows from the commutative property of addition that

$$6 + 5 = 5 + 6$$
$$17 + 35 = 35 + 17$$
$$18,214 + 5,003 = 5,003 + 18,214.$$

Please note that we write equations between arithmetic expressions without evaluating those expressions.

4.3.1.1 Algebraic formulation

To express a mathematical law such as the commutative property of addition, it is common to use **variables**. A variable is a symbol (usually a Latin letter) that may be substituted by any number (it is called a variable because its value may vary). Variables help us express general statements, such as *the sum of every two numbers is independent of their order*.

Similar to **arithmetic expressions**, which express mathematical operations between numbers, **algebraic expressions** express mathematical operations between variables, or between variables and numbers. An equation that equates algebraic expressions or equates an algebraic expression and an arithmetic expression is called an **algebraic equation**.

In its algebraic formulation, the commutative property asserts that for any two numbers, which we shall represent by the variables a and b, the following equation holds:

$$a + b = b + a.$$

This equation states that for *any* two numbers that we may substitute for the letters a and b, the arithmetic expression $a + b$ equals the arithmetic expression $b + a$.

Please note that when we write an algebraic equation that includes a variable that appears multiple times, we must substitute the same number in all its appearances.

4.3.1.2 Commutative property in addition machines

Another way of illustrating the commutative property is by means of the addition machine introduced in Chapter 3 (p. 25). In this context, the commutative property implies that the output of the machine does not change when we insert the same two numbers in reverse order.

For example, the two cases illustrated below result in the same output:

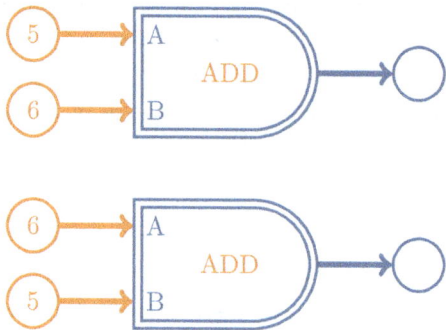

4.3.2 The associative property

Consider the following question:

> There are 3 classes in second grade at our school. There are 24 children in 2A, 28 children in 2B, and 26 children in 2C. How many children are there in second grade?

This problem differs from word problems we have seen so far. All former examples described the union of *two* sets. This is the first time we encounter a problem where *three* sets are joined together. On the surface, there should not be any problem: the union of three sets is as natural as the union of two sets, and it would then seem very simple to define addition for three addends or more. But it turns out that what may be natural in everyday life is not necessarily natural when we want to define an arithmetic operation. For technical reasons, mathematics defines addition as a **binary operation**—an operation that takes only two numbers at a time.

The question is: If addition is only defined for two addends, how can we model a word problem that involves the union of three second grade classes?

In order to solve a problem that involves the union of more than two sets, the union has to be performed in stages, each one involving the union of two sets. For example, we can take the 24 children from 2A and the 28 children from 2B and combine them into one set, and then append the 26 children from 2C to this set. We have thus performed two consecutive summations. The arithmetic expression for these operations is:

$$(24 + 28) + 26.$$

The parentheses indicate the order of operations: first we added 24 and 28, and then we appended 26 to the resulting sum.

The decision to add the children from 2A and 2B first was absolutely arbitrary. We could just as well have first joined together the 28 children from 2B and the 26 children from 2C, and appended the resulting compound set to the 24 children from 2A. The arithmetic expression for that is:

$$24 + (28 + 26).$$

The associative property of addition asserts that in both cases the resulting sum is the same (in short, we say that *addition is associative*). The associative property, like the commutative property, stems from the fundamental fact that the result of counting is independent of the order of enumeration.

In algebraic notation, the associative property asserts that for any three numbers, which we represent by the variables a, b and c:

$$(a + b) + c = a + (b + c).$$

4.3.2.1 Associative property versus commutative property

Many tend to confuse the associative property and the commutative property, thinking that both properties are the same. While both properties stem from the same underlying principle (irrelevance of the order of counting), they are inherently different: the commutative property concerns the addition of *two* numbers, whereas the associative property concerns the addition of *three* numbers.

Let's demonstrate the difference between the commutative property and the associative property by means of addition machines. To add three numbers requires two such machines. A possible arrangement of two addition machines to add the number of children in the second grade classes is presented below:

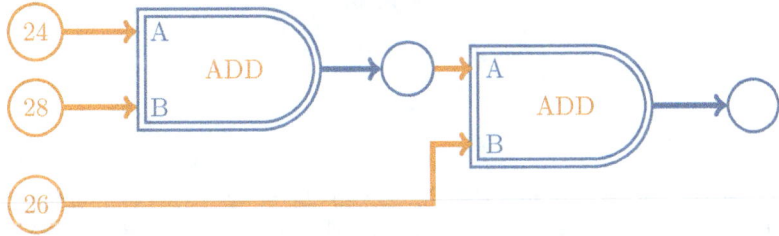

This illustration illustrates the arithmetic expression:

$$(24 + 28) + 26.$$

The commutative property implies that the sum does not change if we invert the order of the addends. Since the above figure features two addition machines, we can invert the order of the addends in either one of them, thus obtaining three new configurations:

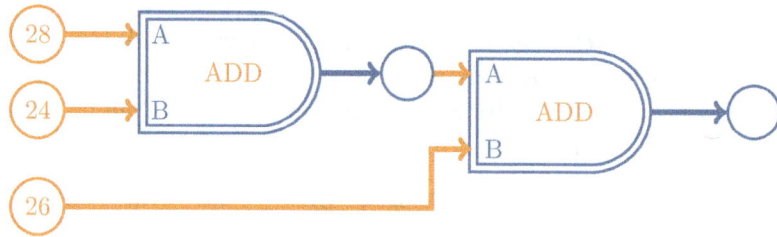

and also:

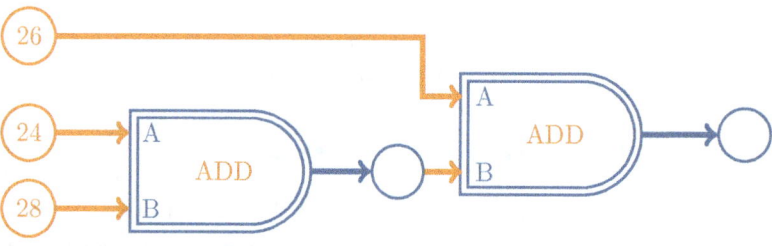

and also:

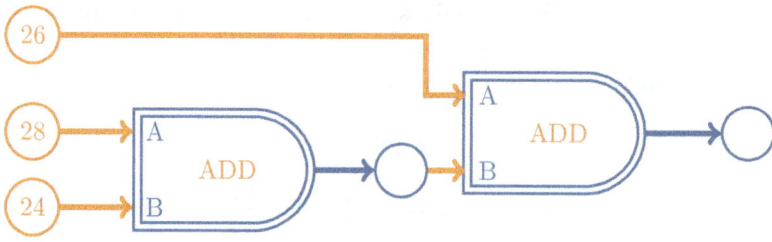

The commutative property implies that in each of the above four configurations the final output is the same. Please note that in all cases the numbers 24 and 28 are added *before* their sum is appended to 26.

The commutative property, however, does not justify the passage from any of the above configurations to the following configuration:

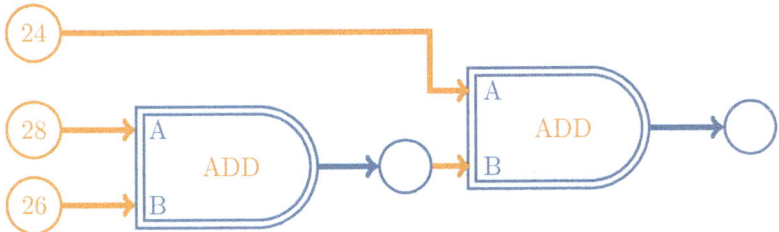

Here 28 and 26 are added before appending their sum to 24. The fact that the final result in the first four configurations is the same as the final result in the last configuration is the essence of the associative property of addition.

4.3.2.2 Combining the commutative and the associative properties

The combination of the commutative property and the associative property implies that when summing up several addends, regardless of the number, the sum is independent of the order in which those addends are joined together.

This fact is useful in computations. For example, when facing the evaluation of the arithmetic expression

$$9 + 7 + 4 + 1 + 3,$$

we can sum those addends in the following order:

$$(9 + 1) + (7 + 3) + 4.$$

The latter expression is easier to evaluate after we have learned to identify pairs of numbers that add up to ten. This example is a first instance of evaluation strategies based on properties of operations.

4.3.3 The identity property of zero

The number zero has a special standing in the context of addition. Consider for example the arithmetic expression

$$9 + 0.$$

It can be interpreted as the size of a set obtained by appending a set that comprises no items (an empty set) to a set that comprises 9 items. Appending an empty set is the same as not doing anything at all. As a result, the compound set has the same number of items as the first set, namely, 9 items.

This example demonstrates a general rule. When one of the addends is zero, the sum is equal to the other addend. For example:

$$23 + 0 = 23$$
$$0 + 346 = 346$$
$$32,171 + 0 = 32,171.$$

This property of zero is called the **identity property** (in mathematics an **identity operation** is one that leaves the object it operates upon unchanged).

Just as we did for both commutative and associative properties, we use algebraic notations to formulate the identity property of zero. For every number, which we represent by the variable a, the following equations hold:

$$a + 0 = a \qquad \text{and} \qquad 0 + a = a.$$

4.3.4 Laws of variation

Suppose we were given that

$$242 + 767 = 1009.$$

Can we exploit this knowledge to easily evaluate the following sums:

$$242 + 769 \qquad \text{or} \qquad 245 + 767?$$

Let's approach this question using a property of addition we are already familiar with—the associative property. Consider the two arithmetic expressions,

$$\text{Sum A} = 242 + 767 \qquad \text{and} \qquad \text{Sum B} = 242 + 769.$$

The first addend in both expressions is identical. However, the second addend in Sum B (the one we want to evaluate) is greater than the second addend in Sum A (the one whose value we are given) by 2. Thus, it is appealing to rewrite Sum B as follows,

$$\text{Sum B} = 242 + (767 + 2),$$

emphasizing that 769 is 2 more than 767.

Now is the moment to apply the associative property of addition. Sum B is a sum of three addends, which we can also evaluate in the following order:

$$\text{Sum B} = (242 + 767) + 2.$$

The expression enclosed in the parentheses is precisely Sum A, hence,

$$\text{Sum B} = \text{Sum A} + 2.$$

This example demonstrates the following fact: if we increase one of the addends and leave the other addend unchanged, the sum increases by the same amount as the addend that was changed. Similarly, if we decrease one of the addends and leave the other unchanged, the sum decreases by the same amount as the addend that was changed. Both rules describe how the sum varies as a consequence of varying the addends. Even though these rules follow directly from the associative property, they highlight a different perspective of associativity, and therefore merit the distinctive name—**laws of variation**.

The combination of both rules leads to the following conclusion: if we increase one of the addends and decrease the other addend *by the same amount*, the sum does not change. This property is often used to simplify calculations. For example, the value of the arithmetic expression

$$29 + 15$$

is the same as the value of the arithmetic expression obtained by increasing the first addend by 1 and decreasing the second addend by 1:

$$30 + 14.$$

The latter expression is easier to evaluate.

4.4 Evaluating addition

Now that we have seen in which situations numbers should be added, and we are acquainted with the properties of addition, we will consider the evaluation of addition. We will see that all evaluation methods, except for learning addition facts by heart, are based on interpretations and properties of addition.

In this chapter, we focus on addition within the range of 20, which, in many countries, is the range that first graders are expected to master. In Chapter 10, we consider the evaluation of sums for arbitrarily large addends.

4.4.1 If you can count, you can add

Consider the following arithmetic expression:

$$7 + 2.$$

The most primitive evaluation strategies are based on counting, as we demonstrate below.

4.4.1.1 Using manipulatives

Assuming we have learned to count, the primitive evaluation strategy builds upon the fundamental interpretation of addition as the result of joining together two sets. To evaluate the sum $7 + 2$, we can perform the following four steps:

(a) Create two sets, one comprising 7 objects and one comprising 2 objects.
(b) Join the two sets together.
(c) Count the objects in the compound set.
(d) The result of the counting is the value of the given arithmetic expression.

We already know the rule: it doesn't matter what the objects are and in what order we count them. This may not be the most effective way to evaluate a sum, but this is how addition should be performed at early age.

4.4.1.2 Using illustrations

An important phase in children's mathematical development is the replacement of material manipulatives by illustrations. Just as it does not matter what the material objects are, it also does not matter how you represent objects in a drawing. The goal is to choose a means of representation that involves a minimal amount of work and yields maximum clarity.

The figure below represents the arithmetic expression 2 + 7:

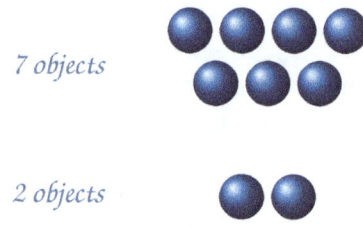

All that remains to be done is to count the total number of objects in both sets. Finding a solution in this way, similar to a solution involving material manipulatives, highlights the connection between addition and counting.

4.4.1.3 Forward counting

Consider the same arithmetic expression, 7 + 2, but this time we interpret addition as **appending**, or **adding to**. For example, we may ask ourselves how many children are in the playground if 2 children joined 7 children who were already there. There is no need to recount the 7 children, but only to account for the 2 newcomers by continuing the number sequence: 8 and 9. Counting that does not start from zero is called **forward counting**, or **counting on**.

4.4.2 Addends whose sum is 10

Because of the special standing of the number 10 in our numeral system, it is essential to have a good command of expressions that sum up to 10. In particular, children should master as soon as possible the following identities:

$$
\begin{array}{ll}
10 + 0 = 10 & 0 + 10 = 10 \\
9 + 1 = 10 & 1 + 9 = 10 \\
8 + 2 = 10 & 2 + 8 = 10 \\
7 + 3 = 10 & 3 + 7 = 10 \\
6 + 4 = 10 & 4 + 6 = 10 \\
5 + 5 = 10 & 5 + 5 = 10
\end{array}
$$

Mastering these identities should work both ways. Children should not only learn to evaluate expressions whose sum is 10, but also to decompose a ten into two addends in all possible ways.

4.4.3 Addition within 20

In many countries, children are expected to be proficient in addition within a range of 20 by the end of the first grade. The range of 20 was not chosen arbitrary; there are two main reasons for insisting on this particular range:

(1) The addition of multi-digit numbers is based on a sequence of steps, each requiring the addition of numbers within the range of 20. Thus, fluency in the latter is required before treating the addition of multi-digit numbers.

(2) Addition within 20 requires children to combine evaluation strategies and place-value considerations. The combination of these two skills comes into action for the first time when adding two single-digit numbers whose sum is a two-digit number.

4.4.3.1 Crossing the first ten

The addition of two single-digit numbers whose sum is a two digit number is termed **crossing the first ten**. To demonstrate why crossing the first ten is so important, let's represent numbers using the fictitious numeral system introduced in Chapter 2, p. 16. The number sequence from 1 to 20 is represented as follows:

| || ||| |||| ||||| ...   • •| ...   ••

Consider the arithmetic expression 3 + 4, which in our fictitious numeral system looks as follows:

$$||| + ||||$$

The evaluation of this expression entails only the removal of the plus sign and placing the two sets of symbols closer together:

$$||| + |||| = |||||||$$

The set of symbols on the right-hand side is the representation of the number 7. As you may have realized, we "evaluated" addition without really doing anything.

In contrast, consider the arithmetic expression 7 + 8. In our fictitious numeral system this expression looks like:

$$||||||| + ||||||||$$

If we remove the plus sign and squeeze the symbols together as shown above, we obtain

$$||||||| + |||||||| = |||||||||||||||$$

The right-hand side, however, is not a valid representation of a number. According to our convention, every ten | have to be replaced by a •, hence,

$$||||||| + |||||||| = •|||||$$

Please note that this evaluation that "crosses ten" consists of two distinct steps: (i) application of the operation, which in this case consists of grouping together the two sets of symbols, and (ii) **regrouping**, which consists of converting decimal units.

The crossing of the first ten is a cornerstone in children's mathematical development, and it is essential for children to fully understand it. Teaching that 7 + 8 equals 15 as a mere "addition fact", without stressing the interplay between evaluation and place-value, misses a key teaching opportunity. Children's first encounter with the evaluation of expressions such as 7 + 8 should be supported by concrete demonstrations, such as the following:

(a) Create two sets of objects (e.g., straws), one comprising 7 objects and one comprising 8 objects.
(b) Join the two sets together.
(c) Group together 10 objects (e.g., wrap a rubber band around 10 straws).
(d) Count separately the number of bundled tens (in this case, 1) and the number of loose objects (in this case 5).
(e) Write down the sum,

$$7 + 8 = 15,$$

emphasizing the meaning of the two-digit number representation.

4.4.4 Using the properties of addition

We have already seen two uses of properties of addition in the evaluation of sums. Two other uses of properties of addition are given here:

(1) Consider the sum $2 + 13$. Children may find it useful to apply the commutative property of addition and prefer to evaluate the sum $13 + 2$. The latter expression calls for **forward counting**, which many children are proficient at.

(2) Consider the sum $14 + 5$. Expressing the first addend as a sum of decimal units, namely, $14 = 10 + 4$, and using the associative property, we may rewrite it as

$$10 + (4 + 5).$$

The sum enclosed in parentheses is within the first ten. The sum of a ten and 9 ones is in fact the definition of the numeral 19.

The use of properties of arithmetic operations is an important skill which underlies all the evaluation algorithms. Like all other skills, it develops gradually. In this context, it is important to let children develop their own solution strategies, and have them compare these strategies. The more strategies children are exposed to, the better they are prepared to address more advanced problems.

4.4.5 The addition table

All the addition expressions in which both addends are within the first ten can be arranged in a table known as the **addition table**.

The addition table is a tool for evaluating sums. For example, in order to evaluate $3 + 8$, one has to draw a vertical line along the column headed by the number 3, and a horizontal line along the row headed by the number 8. The meeting point of the two lines indicates the sum—in this case, 11.

The addition table has several pedagogical values. First, it displays a way to organize data in a resourceful manner. Data organization is a mathematical skill in its own right. Second, the addition table features various patterns whose discovery and elucidation contribute directly to the understanding of addition and its properties. For example, the following properties of the addition table can be addressed:

(1) The first column is identical to the first row, the second column is identical to the second row, etc. In mathematical jargon, the table is **symmetric** with respect to transposition.
(2) In every column and in every row, the numbers appear in sequential order.
(3) In every top-left to bottom-right diagonal, the difference between succeeding numbers is 2.
(4) In every top-right to bottom-left diagonal, the numbers are fixed.

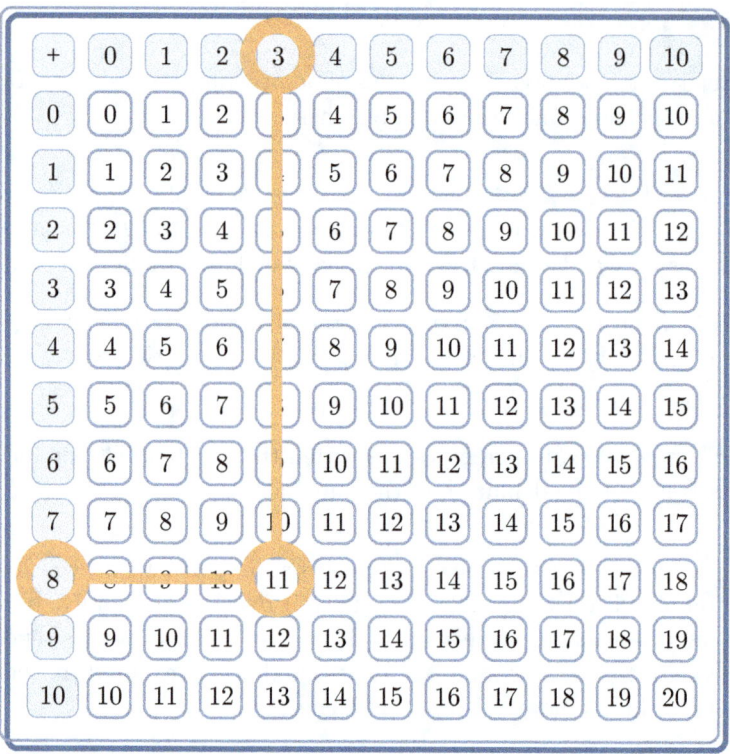

(5) The largest number in the table is located at the bottom right end corner of the table.
(6) Every number in the table features all its possible decompositions into pairs of single-digit numbers.

4.4.6 Addition on the number line

The natural numbers are represented in a number line diagram as equally spaced points on a line. Consider the expression 3 + 4: its number line representation is shown below:

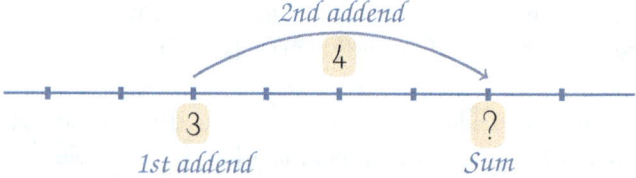

Addition is represented by an arrow. The tail of the arrow points at the first addend, in this case 3. The length of the arrow represents the second addend, in this case 4. The head of the arrow points at the sum.

Please note that this visualization represents addition as **adding to**. As a result, the commutative property is not evident, since the two addends play different roles. This fact is related to our former observation (p. 9) that the number line clearly illustrates the order relation between numbers, but blurs their role as quantifying the size of sets.

Mathematical problems

Problem 4.1. What is the most frequent sum obtained when throwing two dice? What is the most frequent sum obtained when throwing three dice? Explain.

Problem 4.2. How does the commutative property of addition manifest itself in the addition table?

Pedagogical problems

Problem 4.3. While she was teaching addition, Teacher Jane presented the following question to her class:

> Today is Friday. What day will it be 5 days from today?

Gili: This is not a mathematical question!
Alon: This is a mathematical question, however not related to addition.
Tal: This question is related to addition but it cannot be represented in arithmetic form.
Anat: Since Friday is the 6th day, then the answer is 6 + 5 = 11.

(a) What is your opinion regarding the four answers?
(b) In your opinion, should such a question be presented when learning addition?

Chapter 5

Subtraction

Subtraction is the second arithmetic operation learned in school. Subtraction has several interpretations, derived from its inverse relationship to addition. In this chapter, we get acquainted with subtraction: its interpretations and its properties. As in the previous chapter, we address the evaluation of the operation within a number range expected from first graders. The evaluation of subtraction for arbitrarily large operands is addressed in Chapter 11.

5.1 What is subtraction?

Subtraction can be interpreted as a model for four different operations: **taking away** (or removal), **taking apart** (or partitioning), **completion** and **comparison**. Take a look at how different interpretations affect our perception of an operation: consider the following expression:

$$485 - 2.$$

You know what its value is: 483. Now try to list the steps you took: you most likely **took away**, or **removed** 2 from 485. Differently stated, starting off at 485 on an imaginary number line, you took 2 steps backward.

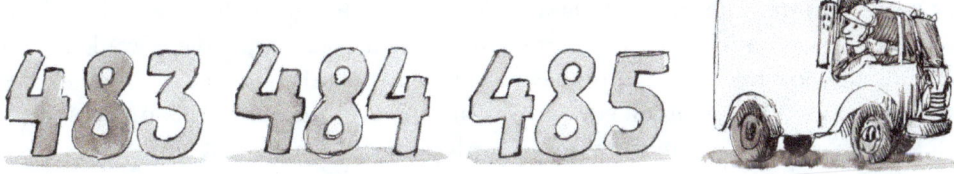

Now, for comparison, consider the following expression:

$$485 - 483.$$

You know what its value is: 2. Try to list the steps you took this time: You probably didn't start out at 485 on an imaginary number line and take 483 steps backward. More likely you asked yourself how much you should *add* to 483 in order to get 485, or what is the *distance* between the two numbers on the number line. This solution strategy reflects a completely different line of thought than the first strategy, and yet in both cases you evaluated subtraction.

5.1.1 Subtraction as a model for taking away

Consider the following word problem:

> Tal had 9 muffins. He shared 3 of them with his friends. How many muffins does he have left?

In this problem, we are given an initial quantity—9 muffins—from which a certain part—3 muffins—was taken away. We have to find out what the remaining quantity is. If you ask your friends to make up a word problem involving subtraction, most of them will likely describe a similar scenario. A typical graphic illustration of this word problem is shown in the figure below:

Let's consider the same word problem from a slightly different angle: the initial quantity, the 9 muffins, can be partitioned into two subsets: the 3 muffins that were given away to friends and the muffins left for Tal. Since the number of the latter is not known, for now let's denote it by a question mark. Thus, the muffins left for Tal, whose number we do not know, plus the 3 muffins given away, add up to the 9 muffins that Tal had originally. We face a situation in which there are two sets that compose a whole; we know one addend, we know the sum, and we have to find out the other addend. We can express this in the form of an addition equation in which one of the addends is **unknown** (we call such an equation an **unknown-addend equation**):

$$\boxed{?} + 3 = 9.$$

This interpretation of the *same* word problem can be illustrated as below:

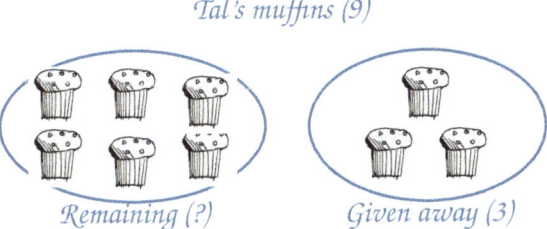

Tal's muffins (9)
Remaining (?)   *Given away (3)*

Here, rather than eliminating the items that were taken away, we separated them from the remaining items. This figure highlights the fact that the original set comprises two subsets—the items that were taken away and the remaining items.

5.1.2 Subtraction as a model for taking apart

Consider the following word problem:

> 12 children are playing in the yard. 7 of them are boys. How many girls are playing in the yard?

In this problem, there is a set of children that comprises two subsets: the boys and the girls. We know that the total number of children is 12, of which 7 are boys. The number of girls is unknown. Once again we denote the unknown quantity by a question mark. Thus, we have to solve the unknown-addend equation:

$$7 + \boxed{?} = 12.$$

We already know that it does not matter whether we count the boys first or the girls first. Therefore, we could just as well reverse the order of the addends:

$$\boxed{?} + 7 = 12.$$

An appropriate illustration of the problem may look as shown below:

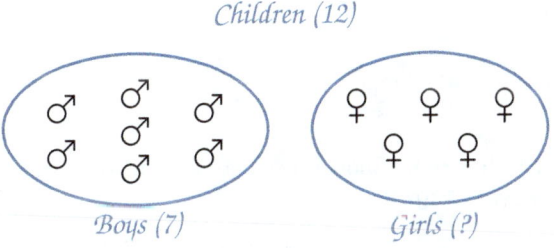

Children (12)
Boys (7)   *Girls (?)*

Please note that in this problem nothing has been taken away. The whole (the 12 children) has been **taken apart**, or **partitioned** into two subsets (boys and girls), the size of one of which (the boys) we know. The task here is to find out the size of the other part (the girls). This turns out to be *exactly* the same task as when part of the whole has been taken away.

5.1.3 Subtraction as a model for comparison

Consider the following word problem:

> Alon has 16 stamps. Gili has 9 stamps. How many more stamps does Alon have than to Gili?

This problem can also be presented as an unknown-addend equation: We know that Alon has more stamps than Gili. If he wished to, he could have partitioned his stamps into two subsets: one set of stamps that is the same size as Gili's (i.e. 9 stamps), and another set the size of which (unknown) is the difference between his number of stamps and hers. Once again, we denote the unknown quantity by a question mark. Now, we have to solve the unknown-addend equation:

$$\boxed{?} + 9 = 16 \quad \text{or} \quad 9 + \boxed{?} = 16.$$

An appropriate illustration of this word problem may look like this:

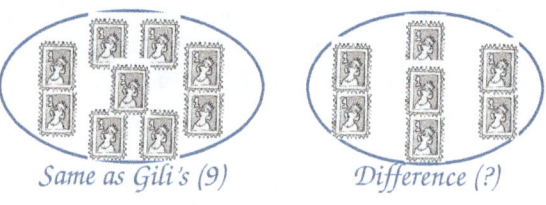

Alon's collection (16)

Same as Gili's (9)   Difference (?)

Please note that the question *"how many more stamps does Alon have?"* can be replaced by the equivalent question *"how many fewer stamps does Gili have?"*.

5.1.4 Subtraction as a model for complementation

Consider the following word problem:

> Anat has \$36. How much money needs to be added so that she can buy a jacket that costs \$49?

Here we take apart the jacket's cost of \$49 into two components: the \$36 that Anat has, and the unknown amount of money that she lacks. We denote the latter amount by a question mark. Thus, we have to solve the unknown-addend equation:

$$\boxed{?} + 36 = 49 \quad \text{or} \quad 36 + \boxed{?} = 49.$$

Try to make your own illustration for this problem!

 Interpretations of subtraction

(1) **Subtraction is the operation inverse of addition**: We introduced the operation of subtraction through four representative word problems. In each case, we ended up with an unknown-addend equation. This is not a coincidence. Subtraction is the inverse operation of addition, in the sense that the known and the unknown are interchanged.

(2) **Beware of cue words**: Look again at the last word problem. We asked: "*How much money needs to be added so that...*". Children tend to identify automatically the word *add* as an indication that they should add some numbers; some end up calculating the sum of whatever numbers are given in the problem.

(3) **Unknowns**: The introduction of equations with an unknown in elementary school is important preparation for middle school algebra.

(4) **Addition-subtraction versus multiplication-division**: The connection between addition and subtraction is akin to the connection between multiplication and division. We will have more to say about these connections in Chapter 8.

5.1.5 The subtraction operation

Now that we have seen a variety of situations in which subtraction is used, we can formally define the operation. The subtraction sign is called **minus**, and it is denoted by −. The first operand is called the **minuend** (a quantity from which another is taken away) and the second operand is called the **subtrahend** (a quantity taken away from another). The result of the operation is called the **difference**.

Consider, for example, the arithmetic expression:
$$9 - 3.$$

Its value is defined as the unknown that solves either of the unknown-addend equations:
$$3 + \boxed{?} = 9 \quad \text{or} \quad \boxed{?} + 3 = 9.$$

Thus,
$$9 - 3 = \boxed{6}$$

because
$$3 + \boxed{6} = 9 \quad \text{or because} \quad \boxed{6} + 3 = 9.$$

In algebraic notation, subtraction is defined as follows: For any two numbers which we denote by a and b, the algebraic expression:

$$a - b$$

is equal to the unknown that solves either of the following algebraic equations:

$$b + \boxed{?} = a \qquad \text{or} \qquad \boxed{?} + b = a.$$

5.1.6 Subtraction on the number line

The connection between subtraction and unknown-addend equations can be illustrated on a number line diagram. Let's first recall how addition is represented on a number line. For example, when we add 6 and 3, there are two possible number line representations:

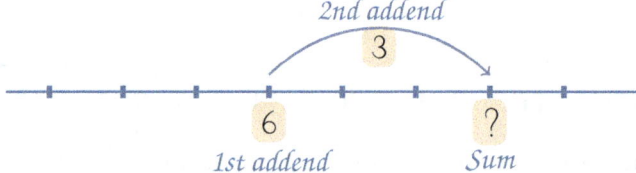

or

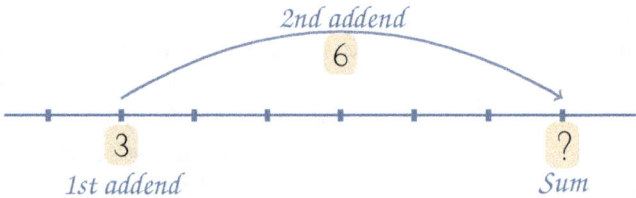

depending on which of the two addends is first.

Consider now the arithmetic expression

$$9 - 3.$$

If we interpret its value as the solution to an equation in which the unknown addend is the *first* operand, namely,

$$\boxed{?} + 3 = 9,$$

then the number line representation looks as shown below.

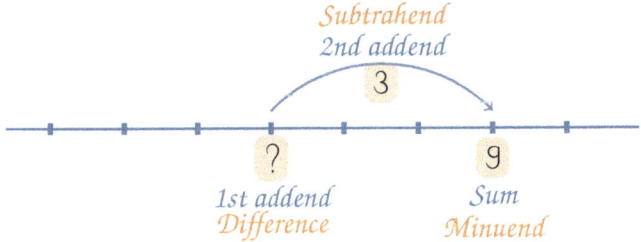

The arrow's tail points to the first addend, which is unknown in this case. The arrow's length represents the second addend, 3. The arrow's head points to the sum, 9. In the transition from subtraction to an unknown-addend equation, the operands change as follows:

Minuend → Sum   Subtrahend → 2nd Addend   Difference → 1st Addend.

If we interpret the value of 9 − 3 as the solution to an equation in which the unknown addend is the *second* operand, namely,

$$3 + \boxed{?} = 9,$$

then the number line representation looks like this:

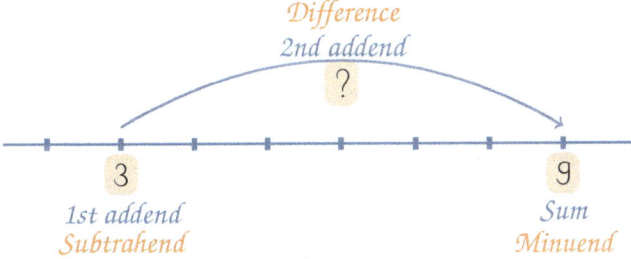

The arrow's tail points to the first addend, 3. The arrow's length represents the second addend, which is now the unknown. The arrow's head points to the sum, 9. In the transition from subtraction to an unknown-addend equation, the operands change as follows:

Minuend → Sum   Subtrahend → 1st Addend   Difference → 2nd Addend

5.2 Word problems

5.2.1 Modeling with subtraction

In the context of subtraction, modeling means identifying those situations in which the solution to the problem is the result of a subtraction operation. Since subtraction has several distinct interpretations, it is necessary that children learn that seemingly very different word problems can be modeled by the same arithmetic expression.

At this point, we can recapitulate the four interpretations of subtraction, with a representative word problem for each interpretation:

(1) **Taking away**: I bought 16 apples and ate 9. How many apples do I have left?
(2) **Taking apart**: In a group of 16 children, 9 are girls. How many boys are in the group?
(3) **Comparison**: Gili has 16 stamps and Alon has 9 stamps. How many more stamps does Gili have?
(4) **Complementation**: Tal has 16 stickers and Anat has 9 stickers. How many stickers should we give Anat for her to have the same number of stickers as Tal?

The solutions to all four word problems is the value of the same arithmetic expression:
$$16 - 9.$$

5.2.2 Interpretation

The freedom to choose an insightful interpretation for a given arithmetic expression turns out to be a real asset when it comes to subtraction. In fact, many of us make such choices on a daily basis without even being aware of it.

Let's revisit the two arithmetic expressions considered on p. 49:
$$485 - 2 \quad \text{and} \quad 485 - 483.$$
Now that we have acquired some terminology, we can explain what it is that guides our choices of interpretation. It turns out that people find it easier to visualize subtraction as a difference between two numbers, the smaller this difference is. In the first expression, the difference is small if we ask which number differs from 485 by 2. This interpretation corresponds to taking away. In the second expression, the difference is small if we ask what is the difference between 485 and 483. This interpretation corresponds to comparison, or complementation.

This leads to the following advice:

 Important!

As part of learning how to *model with mathematics*, children are often asked to write down the equation that *best* matches a given word problem. Consider the question once again:

> Tal has 16 stickers and Anat has 9 stickers. How many stickers should we give Anat for her to have the same number of stickers as Tal?

In your opinion, which of the following equations matches it best:

$$16 - 9 = \boxed{7} \quad \text{or} \quad 9 + \boxed{7} = 16?$$

Both equation match the given word problem equally well; they simply represent different interpretations of subtraction. If we want to educate independent creative thinkers, it is essential not to limit the way they express themselves.

5.3 The properties of subtraction

Throughout this chapter, we have emphasized the reciprocity between addition and subtraction. This relationship also manifests itself in the different properties of the two operations.

5.3.1 Subtraction is not commutative

In general, if we reverse the minuend and the subtrahend, the result, i.e., the difference, does not remain the same. For example, the expression $6 - 3$ does not have the same value as the expression $3 - 6$ (in elementary school, the latter is not even considered a valid expression). In other words, subtraction does not satisfy the commutative property, or in short, *subtraction is not commutative.*

The algebraic formulation of the lack of commutativity is as follows: there exist two numbers, which we represent by a and b, such that

$$a - b \neq b - a.$$

 Very important!

When we introduced the commutative property, we insisted that it had to apply *to all* pairs of numbers a and b. Here, however, when we formulate the lack of commutativity, we are content with the existence of *one* pair of numbers for which commutativity fails. This is because the negation of *"something always holds"* is *"there exists at least one instance in which it does not hold"*.

Please note that if a and b are the same number, then

$$a - b = b - a.$$

And yet, subtraction is not commutative, because we found one case, $a = 6$ and $b = 3$, where $a - b \neq b - a$. (Convince yourself that $a - b = b - a$ *only* if a and b are equal.)

5.3.2 Subtraction is not associative

When adding three numbers, it does not matter in what order we group them (Chapter 4, p. 36). For example, to calculate the sum of 7, 5 and 1, we can choose this order of operations:

$$(7+5)+1,$$

or that order of operations:

$$7+(5+1),$$

and in both cases we obtain the same sum—addition is **associative**.

In an arithmetic expression that has two consecutive subtractions, the order of operations matters. In other words, the value of this expression depends on which operation is performed first. Consider, for example, the following arithmetic expression:

$$10-5-2.$$

If we first subtract 5 from 10, and then subtract 2 from this difference, we obtain:

$$\underbrace{(10-5)}_{5}-2=3.$$

If, however, we first subtract 2 from 5, and then subtract this difference from 10, we obtain:

$$10-\underbrace{(5-2)}_{3}=7.$$

Since all arithmetic expressions must have a well-defined value, there must be a convention to determine the order in which two consecutive subtractions should be performed. The convention is the following: In the absence of parentheses, the operations should be performed in the order in which they appear, from left to right. In the above example,

$$10-5-2=3.$$

In algebraic notation, the non-associativity of subtraction implies that there exist three numbers, a, b and c, for which

$$(a-b)-c \neq a-(b-c).$$

5.3.3 Laws of variation

In this section, we examine how variations in both minuend and subtrahend affect the difference.

5.3.3.1 Varying the minuend

Suppose we were given that

$$841 - 337 = 504.$$

Can we exploit this knowledge to easily evaluate the following differences:

$$844 - 337 \quad \text{or} \quad 839 - 337?$$

Let's approach this question by interpreting subtraction as taking away. Consider the two arithmetic expressions

$$\text{Difference A} = 841 - 337 \quad \text{and} \quad \text{Difference B} = 844 - 337.$$

The minuend, which represents the initial quantity, is larger in B than in A. The subtrahend, which represents the quantity taken away, is the same in both expressions. *The more there was initially, the more remains.* Thus, the difference, which represents the remaining quantity, is larger in B than in A by the same amount as the difference between the minuends.

Since $844 = 841 + 3$, it follows that

$$\text{Difference B} = \text{Difference A} + 3,$$

and since we are given that Difference $A = 504$, it follows that Difference $B = 507$.

Similarly, consider the two arithmetic expressions

$$\text{Difference A} = 841 - 337 \quad \text{and} \quad \text{Difference C} = 839 - 337.$$

This time, the minuend is smaller in C than in A, whereas the subtrahend is the same. *The less there was initially, the less remains.* As a result, Difference C is smaller than Difference A by the same amount as the difference between the minuends. That is, since $839 = 841 - 2$, it follows that

$$\text{Difference C} = \text{Difference A} - 2,$$

and since we are given that Difference $A = 504$, it follows that Difference $C = 502$.

To summarize, subtraction satisfies the following laws of variation:

(a) *Increasing the minuend increases the difference by the same amount.*
(b) *Decreasing the minuend decreases the difference by the same amount.*

These laws of variation can be illustrated on the number line as shown below:

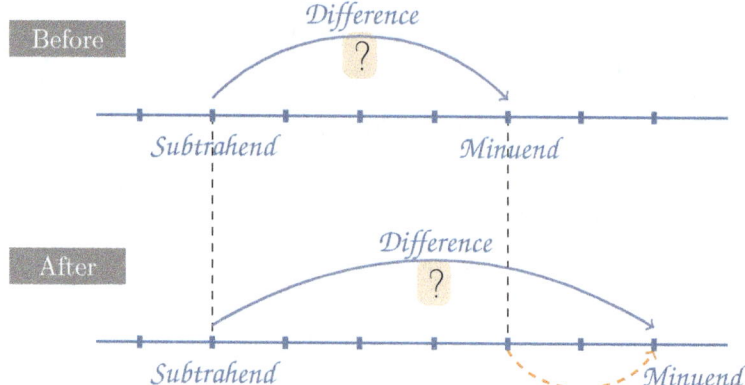

Increasing the minuend (i.e. moving the arrow's head to the right) without changing the subtrahend (i.e. keeping the arrow's tail fixed) increases the difference (i.e. increases the arrow's length). Likewise, we can conclude that *decreasing the minuend decreases the difference by the same amount.*

5.3.3.2 Varying the Subtrahend

And now, suppose we were given that

$$841 - 337 = 504.$$

Can we exploit this knowledge to easily evaluate the following differences:

$$841 - 339 \quad \text{or} \quad 841 - 334?$$

This time, we varied the subtrahend. If we stick to the interpretation of subtraction as taking away, then every child knows the rule: *the more we remove the less remains*, and conversely, *the less we remove the more remains.*

Thus, subtraction also satisfies the following two laws of variation:

(a) *Increasing the subtrahend decreases the difference by the same amount.*
(b) *Decreasing the subtrahend increases the difference by the same amount.*

5.3.3.3 Varying both Minuend and Subtrahend

What should we expect to happen if we varied both minuend and subtrahend by the same amount? That is, the initial amount was larger, but also equally as much was removed. As a result, the difference remains the same, as illustrated in the number line diagram shown below.

Changing both minuend and subtrahend by the same amount (i.e. moving both arrow's tail and head in the same direction and the same distance) does not change the difference (i.e. the arrow's length does not change).

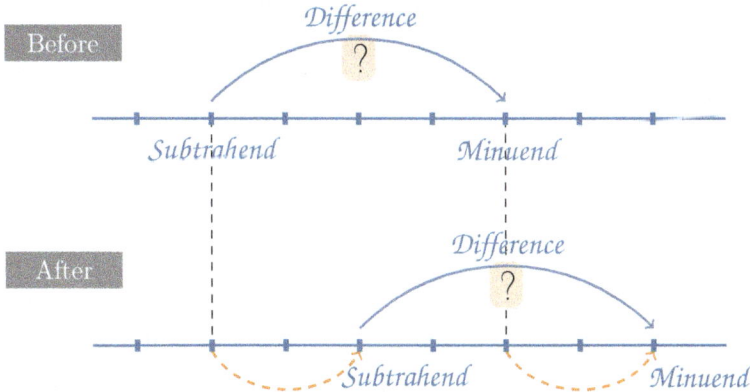

The fact that the difference does not change when both minuend and subtrahend are varied by the same amount may be helpful in evaluating differences. Consider the following evaluation problem:

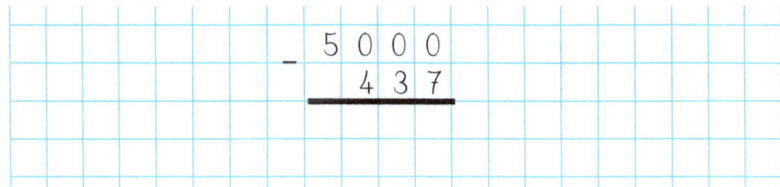

The standard algorithm for solving this problem necessitates a three-step regrouping (see Chapter 11 ahead). Instead of using the standard algorithm, we can exploit the fact that the difference remains the same if we reduce both the minuend and the subtrahend by 1. We therefore replace the given evaluation problem by a *different* evaluation problem having the *same* solution. The latter does not necessitate any regrouping, hence it is much easier to solve:

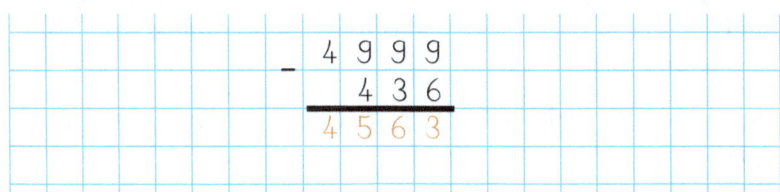

5.3.3.4 Simplifying algebraic expressions

Middle school pupils learn to simplify algebraic expressions. When they are given an algebraic expression and asked to find a simpler equivalent expression (a property not always well-defined which may involve some contrived guesswork) they learn certain rules. One of them is the following: Subtraction before parentheses reverses the sign. What does that mean?

For example, the algebraic expression:
$$a - (b + c)$$
is equivalent to the following algebraic expression:
$$a - b - c.$$
The latter has no parentheses, hence, is in some sense simpler. The removal of the parentheses resulted in turning the plus sign between b and c into a minus sign.

Does this look familiar? This is precisely one of the laws of variation. Middle schoolers memorize the sign reversal rule, being unaware that it stems from a very intuitive fact—*the more you remove the less remains*.

5.3.3.5 Interchanging addition and subtraction

The laws of variation of addition and subtraction combined lead to the following useful rule: when adding and subtracting numbers, changing the order of operations does not change the final result, as long as each number retains the operation that preceded it. For example,
$$13 + 45 - 6 + 12 - 8 - 9 + 14 = 45 + 14 - 8 + 13 - 9 - 6 + 12.$$

5.3.4 Adjoint subtraction equation

Suppose we are given the following identity:
$$673 - 137 = 536.$$
Then, the following identity also holds:
$$673 - 536 = 137.$$
The second equation was derived from the first equation by interchanging the subtrahend and the difference. We say that the second equation is **adjoint** to the first, or that both equations are adjoint to each other.

The first equation implies the second equation because both derive from the same addition equation:
$$536 + 137 = 673.$$
Differently stated, the value of
$$673 - 137$$
is the number that when added to 137 yields 637. If we know that number, 536, then we automatically know the number that when added to 536 yields 637. It is 137.

5.3.5 Subtraction of zero

The number zero also has a distinctive property in subtraction. Taking away zero from any quantity does not actually take away anything. In other words, for any number a:
$$a - 0 = a.$$
Another interpretation of this equation is that the distance on the number line between a and 0 is a. The equation adjoint to $a - 0 = a$ is obtained by interchanging subtrahend and difference:
$$a - a = 0.$$
This equation can also be interpreted in two ways. (1) If you remove a from a, there is nothing left. (2) The distance on the number line between a and itself is 0.

5.4 Evaluating subtraction

Now that we understand in which situations numbers are to be subtracted, and now that we are acquainted with the properties of subtraction, we can consider the question of how to calculate differences.

5.4.1 If you can count, you can subtract

In Chapter 4, we demonstrated how addition is based on counting. Since subtraction is the inverse operation of addition, then it must also be based somehow on counting. To illustrate this, consider the following word problem:

> Granny Smith's family bought a crate of 24 apples. As of this morning, they have eaten 15 of those apples. How many apples are left?

To solve this problem, let us draw 24 symbols, each symbol representing an apple:

||||||||||||||||||||||||

We then cross out 15 symbols, one for every apple that was eaten:

~~||||||||||||||||~~ |||||||||.

We finally count the signs we did not cross out, each representing an apple that was not eaten. This method of calculation may be inefficient, but it always works!

5.4.2 Backward and forward counting

There are two fundamental approaches to evaluating subtraction, each based on a different interpretation of the operation. One, **backward counting**, is based on the interpretation of subtraction as taking away. The other, **forward counting**, is based on the interpretation of subtraction as complementation.

We demonstrate both approaches for the evaluation of the following expression:
$$143 - 56.$$

5.4.2.1 Backward counting

One approach is to ask: how much remains when we take away 56 from 143. Rather than running the number sequence backward 56 steps from 143, we can step backward in longer leaps. For example,

(a) Take 43 away from 143. Remainder: 100.
(b) Only 43 out of 56 were taken away. How much more needs to be taken away? $56 - 43 = 13$.
(c) Take 10 away from 100. Remainder: 90.
(d) Only 10 out of 13 were taken away. How much more needs to be taken away? $13 - 10 = 3$.
(e) Take 3 away from 90. Remainder: 87.
(f) Thus, $143 - 56 = 87$.

Children are frequently asked to write down the sequence of operations that lead them to the final answer. A recurrent, though erroneous, way of writing is as follows:

$$143 - 56 = 143 - 43 = 100 - 10 = 90 - 3 = 87.$$

This is incorrect because arithmetic expressions on both sides of equal signs are not equal. An admissible alternative is:

$$143 - \underbrace{56}_{=43+13} = 100 - \underbrace{13}_{=10+3} = 90 - 3 = 87.$$

5.4.2.2 Forward counting

Another approach is to ask: how much should we add to 56 to obtain 143? Rather than running the number sequence forward from 56 to 143, we can step forward in longer leaps. For example,

(a) Complement 56 to the next ten: add 4 to 56 and get 60.
(b) Complement 60 to the next hundred: add 40 to 60 and get 100.
(c) Add 43 to 100 to get 143.
(d) In total we added: $4 + 40 + 43 = 87$.
(e) Thus, $56 + 87 = 153$, and consequently, $143 - 56 = 87$.

This sequence of operations can be written as a list of equations:

$$56 + 4 = 60$$
$$60 + 40 = 100$$
$$100 + 43 = 143.$$

Hence

$$56 + \underbrace{4 + 40 + 43}_{87} = 143.$$

This approach, which is based on complementing to the next decimal unit, is commonly used in monetary transactions involving change.

Forward counting can be visualized on a **blank number line**. A blank number line is a visual aid that mimics a number line diagram. Points are placed on it on demand, keeping the order relation between numbers right, but not the scale. Starting at the subtrahend, we move along the blank number line until we reach the minuend:

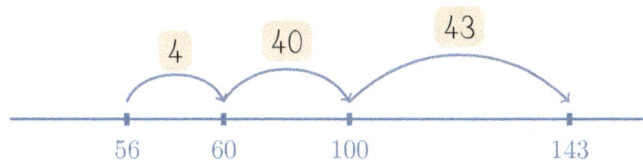

All that remains to do is sum up the increments.

5.4.3 Subtraction within 20

First graders are expected to easily be able to subtract within the range of 20. One of the teacher's goals is to help children develop evaluation strategies that eventually lead to fast and automated calculations. In this context, I strongly recommend accustoming children to interpret subtraction in all possible ways: taking away, taking apart, comparison and complementation. Moreover, subtraction is easier when you understand its relation to addition.

Subtraction within 10 can be practiced using concrete manipulatives, illustrations, and eventually by completing number sentences.

Subtraction within 20, just as addition within 20, requires to combine evaluation strategies and place-value considerations (see Chapter 4, p. 43). Similar to the add-to-ten strategy used in addition, there is a **subtract-to-ten** strategy for subtraction. Consider the following arithmetic expression:

$$14 - 6.$$

The subtract-to-ten strategy works as follows::

(a) Take away 4 from 14. Remains 10 (no regrouping needed).
(b) Only 4 out of 6 were taken away. To find how much more needs to be taken away, we look for a number, which when added to 4 yields 6. That number is 2.
(c) Take away 2 from 10. The answer is 8. This is a subtraction problem within 10, which we should be capable of performing by now.
(d) Thus, $14 - 6 = 8$.

5.4.4 Validation

A very important practice in mathematics is the **validation** of solutions. Validation is a procedure whose purpose is to verify that an answer that was obtained is indeed the solution to the problem. Unfortunately, children are rarely taught this skill. Left to their own devices, children simply re-solve the problem. Then, if they happen to have a systematic error (as opposed to a typographic error), they will likely err again.

An effective way of validating that a subtraction was evaluated correctly is to solve the inverse addition problem. Suppose, for example, that we obtained the following solution:

$$43 - 19 = \boxed{25}.$$

We can validate the solution by verifying whether:

$$\boxed{25} + 19 \stackrel{?}{=} 43.$$

This validation technique can be combined in a meaningful word problem. For example:

> 43 people came to the party, and 19 left. How many people remained?

Children who found that there were 25 people left can then ask themselves how many people will there be at the party if the 19 who left decided to return and join the remaining 25. If they can add properly, they will discover that the answer is 44, rather than 43. Then, they will see that they got it wrong either when they solved the subtraction problem, or when they were validating their answer. In either case, they have to re-examine their solution.

5.5 Negative numbers

In most math curricula, negative numbers are not learned before middle school. Yet, there is no reason to ignore the children's daily encounter with negative numbers when they are much younger. Negative numbers are found, for example, in temperatures below freezing, in the parking level at the shopping mall, or in the altitude above sea level of Death Valley.

Negative numbers are numbers that are less than zero. The number (-2), for example, is defined as the outcome of taking away 2 from 0, i.e., it is defined to be the value of the arithmetic expression:

$$0 - 2.$$

A child who has internalized the laws of addition and subtraction should be prepared to understand the arithmetic of negative numbers as well. For example, knowing that $a - b = c$ implies that $b + c = a$, one could infer that the definition $0 - 2 = (-2)$ implies that

$$2 + (-2) = 0.$$

As another example, consider the arithmetic expression:

$$4 - (-2).$$

Children in middle school are often taught to memorize rules such as *minus-minus makes plus*. However, a child who was taught early on that subtraction may be interpreted as an unknown-addend equation, may be able to evaluate this expression by looking for the unknown in the following equation:

$$(-2) + \boxed{?} = 4.$$

Rephrased that way, even elementary schoolers may know the answer, especially if they use the number line:

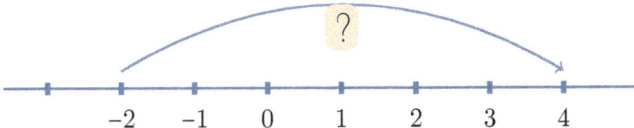

We will not spend more time on negative numbers, however, I want to point out that advanced mathematics often rests on foundations laid in the early grades. Negative numbers are a case in point.

Mathematical problems

Problem 5.1. A bus with passengers leaves the first stop. At the next stop, 10 passengers get on the bus. At the next stop, 1 passenger gets off and 8 passengers get on. Then 2 get off and 5 get on. Then 6 get off and 6 get on. At the next stop, 9 passengers get off and 1 gets on. At this point, all the seats on the bus are taken, and 2 passengers remain standing. The bus has 51 seats. How many passengers initially got on the bus at the first stop?

Problem 5.2. Use the fact that

$$14{,}732 - 8{,}249 = 6{,}483$$

to evaluate the following expressions:

$$14{,}733 - 8{,}249 \qquad 14{,}730 - 8{,}249 \qquad 14{,}732 - 8{,}251$$
$$14{,}732 - 6{,}483 \qquad 15{,}732 - 9{,}249 \qquad 8{,}249 + 6{,}483.$$

Explain which properties of subtraction you used each time.

Problem 5.3. How will the difference in a subtraction problem change if we increase the minuend and decrease the subtrahend by the same amount? Verify your hypothesis on a particular example.

Problem 5.4. Anat's age is exactly 5 years, 3 months and 21 days. How much time will elapse until she be exactly 10 years old? (Let's assume that there are 30 days in a month.) What calculation strategy did you use?

Pedagogical problems

Problem 5.5. Consider the question: *how much should you add to 14 to get 21?* How would you explain to children that they have to subtract, even though it would seem that the operation is one of adding?

Problem 5.6. Find ten word problems in a child's math book that are related to subtraction, and for each one note what interpretation of subtraction is involved. Try to estimate which interpretations are more frequent.

Problem 5.7. Teacher Marius presented his class with the following word problem:

> Gili has $30. She bought a book for $14 and a pencil for $2. How much money does she have left?

Teacher Marius asked each child to write an appropriate arithmetic expression.

Martin wrote: $30 - (14 + 2)$.
Sabina wrote: $30 - 14 - 2$.

(a) Are both expressions equivalent?
(b) Do both expressions correctly represent the given word problem?
(c) Which property of subtraction is illustrated in this question?

Chapter 6

Even and Odd Numbers

The natural numbers can be partitioned into two subsets: **even numbers** and **odd numbers**. The property of a natural number to be either even or odd is called **parity**. Parity is an instance of a more general property of natural numbers called **divisibility**. In fact, the teaching of parity at an early age is a not to be missed opportunity to plant seeds toward the learning of division at a later stage. Children as young as first graders have a fairly good intuitive understanding of what parity means. Rightly handled, this understanding can be consolidated so that the more intricate concept of divisibility can later be related to a capacity that children already possess.

6.1 Definition of parity

Parity is a property of whole numbers, and nothing but whole numbers. We now define what even and odd numbers are in two different ways:

> **Definition 6.1.** A whole number is **even** if it can be represented as the sum of two equal whole numbers. A whole number that cannot be represented as the sum of two equal whole numbers is called **odd**.

According to Definition 6.1, a set contains an even number of items if it can be partitioned into two equally-sized sets. For example, 6 is even because 6 = 3 + 3, but 19 is odd because it cannot be represented as the sum of two equal whole numbers.

> **Definition 6.2.** A whole number is **even** if it can be represented as a multiple of 2. A whole number that cannot be represented as a multiple of 2 is called **odd**.

According to Definition 6.2, a set contains an even number of items if we can arrange all its elements in pairs, without any element left over. Said otherwise, a set contains an even number of items if we can partition it into a collection of sets, each containing 2 items. For example, 14 is an even number because:

$$14 = 2 + 2 + 2 + 2 + 2 + 2 + 2.$$

All mathematical concepts must be defined unambiguously. How can this be reconciled with the fact that we gave two definitions for the parity of a number? Can a number be even according to Definition 6.1, and odd according to Definition 6.2? The answer is no. Both definitions are equivalent, and therefore mathematical consistency is not violated.

Both definitions of parity are equally useful, and therefore both should be taught. Since it is essential not to leave children with the impression that the same concept can be given two unrelated definitions, it is imperative that they understand why both definitions are equivalent. Moreover, the two definitions of parity will recur later in the form of the two interpretations of division—**partitive division** and **quotative division**.

So without further ado, let's show that both definitions are indeed equivalent:

Claim 6.1.

(A) If a natural number is even according to Definition 6.1, then it is also even according to Definition 6.2.

(B) If a natural number is even according to Definition 6.2, then it is also even according to Definition 6.1.

Proof of Part A of the claim

(1) Suppose that a natural number is even according to Definition 6.1.
(2) It follows that this number represents a collection of items that can be divided into two equally-sized sets.
(3) The fact that two sets are of equal size means that the items in both sets can be matched exactly (Chapter 1, p. 2): every item in one set can be matched with a single item from the other set, without any leftover items.
(4) If every item in one set has a match in the other set, it means we have partitioned the original set into pairs.
(5) Since all items can be partitioned into pairs, their number is even according to Definition 6.2, which is precisely what we wanted to show.

The figure below illustrates the transition from equally-sized sets into a collection of pairs:

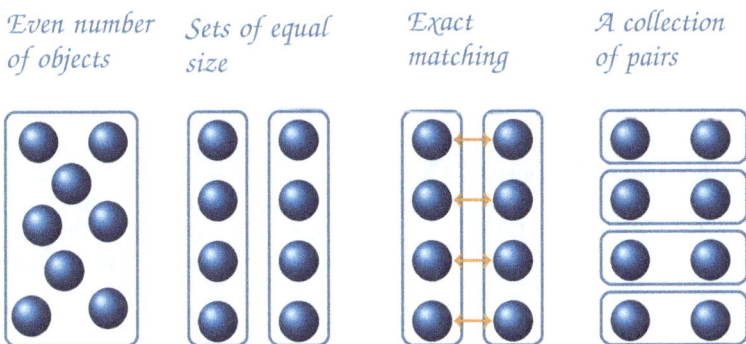

Proof of Part B of the claim

(1) Suppose that a number is even according to Definition 6.2.
(2) It follows that this number represents a collection of items that can be partitioned into pairs.
(3) In every pair of items, we may label one item *first* and the other item *second*.
(4) Take all the items labeled *first* and put them together. Take all the items labeled *second* and put them together.
(5) We have partitioned the collection of items into two equally-sized sets, since every item labeled *first* in one set is matched with an item labeled *second* in the other set.
(6) It follows that the number of items is even according to Definition 6.1, which is precisely what we wanted to show.

To see how the definitions of parity come into play, consider the following question:

Is the total number of feet in the classroom even or odd?

This question can be answered using either definition of parity.

(a) The total number of feet is a sum of pairs, since each child has two feet. Therefore the total number of feet is even.
(b) The set of feet can be partitioned into two subsets—right feet and left feet—and there is an equal number of each. Thus the total number of feet is the sum of two equal whole numbers, and therefore it is even.

Is zero an even number?

Many people are confused by whether zero is even or odd. I have heard more than once the claim that it makes no sense to assign zero a parity.

As always, in case of confusion, go back to definitions. According to Definition 6.1, zero would be even if, and only if, we can represent it as the sum of two equal whole numbers. Since:

$$0 + 0 = 0,$$

it follows that zero is an even number.

 **Activity: Even and odd numbers**

(1) Provide children with a collection of items, for example buttons.
(2) Ask them to determine whether the number of buttons is odd or even, in two different ways—according to Definition 6.1 and according to Definition 6.2.
(3) According to Definition 6.1, they should remove one pair of buttons at a time, either until no button remains (in which case we conclude there was an even number of buttons), or until there is only one button left (in which case we conclude there was an odd number of buttons).
(4) According to Definition 6.2, they should try to separate the buttons into two equally-sized sets. This can be done by removing buttons two at a time and placing each in a different group.
(5) Note how similar both procedures are, which indicates that both definitions of parity are equivalent.

6.2 Arithmetic of parity

Consider the following question:

> There are two crates of oranges. It is known that the number of oranges in each crate is odd. Is the total number of oranges odd or even?

Imagine that pairs of oranges were bagged together. Since there is an odd number of oranges in each crate, each crate must contain bagged pairs of oranges along with a single orange left over. When we merge the contents of the two crates, we get a bunch of bagged pairs of oranges, and two loose oranges, one from each crate. Bagging these two loose oranges together results in all oranges being paired together. Therefore, the total number of oranges in the two crates is even.

Even and Odd Numbers

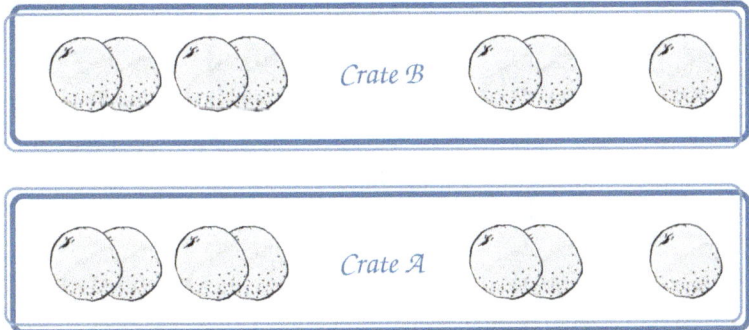

The analysis of this problem along with similar ones leads to the following conclusions:

(a) The sum of two even numbers is always even.
(b) The sum of an even number and an odd number is always odd.
(c) The sum of two odd numbers is always even.

Please note that this is the first time that we formulate a statement about the outcome of an arithmetic operation involving two numbers without knowing what the precise numbers are.

It is possible to summarize the above conclusions in the form of an **addition table of parity**. The table below describes how the parity of the sum depends on the parity of the addends.

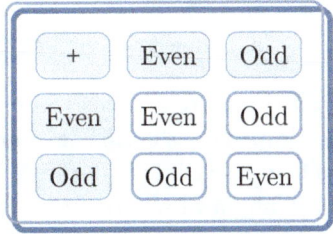

This table is used in a similar way as the addition table presented in Chapter 4. For example, to find out the parity of a sum of two odd numbers we draw a vertical line along the column headed by the word *Odd*, and a horizontal line along the row headed by the word *Odd*. The meeting point of the two lines indicates the parity of the sum—in this case, *Even*.

In a similar manner, we construct a **multiplication table of parity**. This multiplication table shows, for example, that the product of two odd numbers is odd.

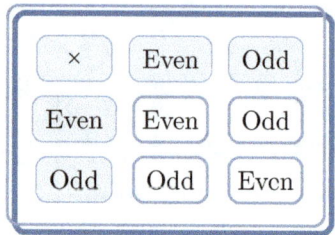

6.3 Determining the parity of a number

How do we know whether a given natural number is odd or even? The rule is to inspect the parity of the ones digit—the parity of the number is the same as the parity of its ones digit. This rule is the children's first encounter with a **divisibility rule**—a simple, easy-to-use rule to determine whether a given natural number is divisible by another given natural number.

Let's then formulate the claim:

> **Claim 6.2.** *A natural number is even if, and only if, its ones digit is even.*

The phrase "if, and only if", which is frequently used in mathematical statements, merits some explanation. When we say that *Property A holds if, and only if, Property B holds*, we make two distinct claims:

(1) *Property A holds if Property B holds* means that if Property B is true then so is Property A (in other words, Property B implies Property A).

(2) *Property A holds only if Property B holds* means that if Property B does not hold then neither does Property A. This in turn means that if Property A holds then Property B must hold as well, for otherwise, Property A wouldn't hold (in other words, Property A implies Property B).

Thus, Claim 6.2 constitutes in fact two distinct claims:

(1) If the ones digit of a natural number is even, then the number is even.
(2) If a natural number is even, then its ones digit is even.

Proof of Claim 6.2

Consider, for example, the number 846. We can prove it is even according to Definition 6.1 by showing that 846 = 423 + 423. This argument would remain specific to this example, and would offer no insight into possible generalizations. A more insightful affirmation that 846 is even is given by the following sequence of arguments:

(1) 10 is even because it is the sum of two equal whole numbers: $10 = 5 + 5$.
(2) The numbers $20, 30, \ldots, 90$ are even because they can be represented as sums of even numbers:
$$20 = 10 + 10, \qquad 30 = 20 + 10, \qquad \ldots, \qquad 90 = 80 + 10.$$
(3) 100 is even because it is the sum of two equal whole numbers: $100 = 50 + 50$.
(4) The numbers $200, 300, \ldots, 900$ are even because they can be represented as sums of even numbers:
$$200 = 100 + 100, \qquad 300 = 200 + 100, \qquad \ldots, \qquad 900 = 800 + 100.$$
(5) 846 is the sum of 800, 40 and 6. Since all three of them are even numbers, their sum is an even number.

If we were rather considering the parity of 847, we would find out that it is the sum of 800, 40 and 7. The sum of two even numbers and an odd number is odd. Generally, hundreds and tens are always even, hence it is only the parity of the ones digit which determines the parity of the entire number.

Please note: we used a very important pedagogical strategy here: *proof by representative examples*. A mathematically sound proof to the Claim 6.2 would require some algebra, and would therefore not be accessible to children before middle school. What we have done is to show why the claim is correct in particular examples, using arguments that are valid for any other number as well.

6.4 More on parity

In this section, we consider two additional topics associated with parity.

6.4.1 Why don't we define parity for fractions?

In the beginning of this chapter, we emphasized that parity is a property of whole numbers only. Why not apply it to fractions as well? We could propose, for example, that a fraction is even if its numerator is even. Thus we would say that
$$\frac{2}{3} \quad \text{and} \quad \frac{4}{11} \quad \text{are even,}$$
whereas
$$\frac{3}{8} \quad \text{and} \quad \frac{5}{78} \quad \text{are odd.}$$
Alternatively, we could have decided that it is the denominator that determines the parity of a fraction. Both definitions, however, are unacceptable because they are ambiguous. The fraction 3/5, for example, which would be considered as odd according to the above criterion, can also be represented as 6/10, which would render it an even number according to the same criterion.

6.4.2 Parity in error control

Computers are machines in which all the information is represented using only two digits: 0 and 1 (also known as **binary digits**, or **bits**). A sequence of bits (usually 16, 32, or 64 bits) is called a **computer word**. When a message is transmitted from one computer (the sending machine) to another (the receiver machine), this information is encoded by a sequence of computer words. Computers are made of man-made electronic devices, and as such, they sometimes do not behave as expected. In particular, errors may occur during the transmission of messages. Since a message is a sequence of bits, an error manifests either as a 0 turning into a 1, or vice-versa, as a 1 turning into a 0. How can the receiver machine detect whether an error has occurred during the transmission of messages?

A common method for checking for errors is based on parity. The probability of an error to occur in a given word is usually quite slim. The probability of two errors occurring in the same word is even smaller, to an extent that we may assume that this never happens. In other words, we can assume that all faulty words have exactly one erroneous bit. This fact can be exploited as follows: in every computer word the last digit is allocated as a **control digit**. It is set by the sending machine so that the total number of 1's in the word is even. For example, in the following computer word, there are eight 1's, hence the sending machine sets the control number to be zero:

$$1110010101010010.$$

If one of the digits in a word is inverted due to an error, whether it is a 0 that turned into a 1 or the other way around, the total number of 1's will become odd.

Suppose now that we get a message in the form of a sequence of computer words. The receiving machine may inspect this message, word by word, checking whether the number of 1's in each word is even. If it comes across a word with an odd number of 1's, it knows right away that an error has occurred in this word during transmission, and generates an error message.

Mathematical problems

Problem 6.1. Alon had a jar full of marbles. In order to find out whether there was an odd number or an even number of marbles, he began to take them out of the jar, two at a time. After a while he got tired, left the jar partially full, and the rest of the marbles all over the rug. The following morning, his sister Gili counted the marbles that were left inside the jar and found that there were 43. Was the initial number of marbles in the jar odd or even? Explain.

Problem 6.2. Suppose that the sum of two numbers is even. Can their difference be odd? Justify your answer.

Problem 6.3. Parity at-a-glance: Consider the four flocks of birds shown below:

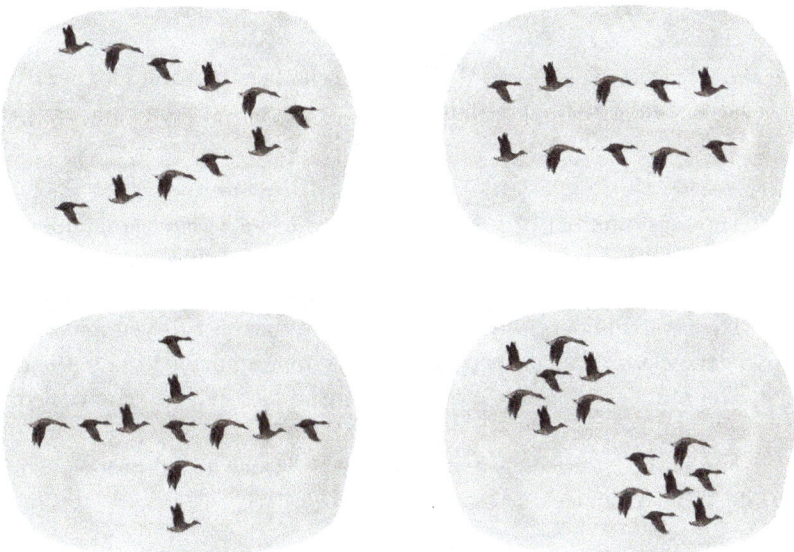

(a) Without counting the birds, determine whether the total number of birds in each flock is even or odd.
(b) Which definition of parity did your answer rely on in each case?
(c) Create at least four new flocks to be used in a *parity at-a-glance game*. With each new flock, try to represent another property of parity (e.g., the addition rules for parity).

Problem 6.4. There are 100 people in a room. Each one of them is wearing a hat. Some of them are wearing white hats and some of them black hats. Each of them can see each hat except their own. Find a short sentence that, said out loud, will enable every person in the room to know at once the color of their own hat (hint: think of the role of the control digit in a computer word).

Pedagogical problems

Problem 6.5. Teacher Kate asked the children whether the total number of feet in her class was even or odd (without peeking under the tables). The children gave several replies:

Chris: The total number of feet is even because it is twice the number of children.
Veronica: The total number of feet is even because every child has two feet. Hence the total number of feet is two plus two plus ... plus two.
Willow: The total number of feet is even because there is an even number of children in the class. (Willow wasn't sure how to account for the teacher's feet.)
Nur: The number of feet is even because the feet can be partitioned into right feet and left feet.
Karen: The number of feet is even because the feet can be partitioned into boys' feet and girls' feet.

(a) Which answers would you classify as correct?
(b) What would you reply to each child?
(c) Which answer you would classify as most accurate? Is there more than one?
(d) Which answer displays best, in your opinion, the notion of parity? Is there more than one?

Chapter 7

Multiplication

Multiplication is the third arithmetic operation learned in school. Its most frequent interpretation is as repeated addition, hence its comprehension relies foremost on a profound understanding of addition. This chapter offers a first acquaintance with multiplication, notably its interpretations, its properties, some applications, and the evaluation of products of single-digit operands. The evaluation of products of multi-digit operands is treated in the second volume of this book.

7.1   What is multiplication?

7.1.1   Multiplication as a model for repeated addition

Consider the following word problem:

> There are 4 boxes, each containing 3 coins. How many coins are there in total?

Clearly, this problem can be modeled using addition. The total number of coins is given by the arithmetic expression:

$$3 + 3 + 3 + 3.$$

This expression features a very distinctive pattern: the same amount is added repeatedly. In this case, 3 is added 4 times. We can say then that the total number of coins is 4 **times** 3.

The above word problem is simple enough so that its solution can be expressed by a chain of additions. But what would we do if there were instead 186 boxes, containing 25 coins each? The total number of coins would then be given by the arithmetic expression:

$$25 + 25 + 25 + \cdots + 25 + 25,$$

where the addend, 25, is added repeatedly 186 times. The writing alone, not to mention the evaluation of the sum, would have taken us quite a long time! Since situations where the same amount in added again and again are very common, a dedicated mathematical operation was invented for this purpose: **multiplication**.

7.1.2 Multiplication as a model for proportional comparison

Multiplication also arises in contexts of comparison between quantities. Consider the following word problem:

> Gili has 6 toy cars. Alon has 4 times as many toy cars as Gili. How many toy cars does Alon have?

In this question, the number of toy cars that Alon has is given in relation to the known number of toy cars that Gili has. We may view this problem as another instance of repeated addition: to know how many cars Alon has we just have to add 4 times the number of cars that Gili has.

Please note that we now have different quantitative comparisons between quantities, each associated with a distinct operation: **4 items more than** (addition), **4 items fewer than** (subtraction), and **4 times as many items as** (multiplication).

7.1.3 The multiplication operation

The shorthand notation for, say, adding 25 repeatedly 186 times is:

$$186 \times 25,$$

or in words, 186 **times** 25. The operation sign × is called the **multiplication sign**. An alternative symbol for multiplication is a dot, that is,

$$186 \cdot 25 = 186 \times 25.$$

Like any other arithmetic operation, multiplication operates on two operands. The first operand (in the above case, 186) is called the **multiplier**, and it represents how many times the same amount is added repeatedly. The second operand (in the above case, 25) is called the **multiplicand**, and it represents the amount that is added repeatedly. In certain situations, we do not distinguish between the two operands, calling both **factors**. The outcome of the operation is called the **product**.

The multiplier and the multiplicand play different roles in multiplication. This distinction is sometimes overlooked because multiplication is commutative: the product does not change when the multiplier and the multiplicand are interchanged (see p. 82 ahead). Yet, for example, the arithmetic expressions 5 × 3 and 3 × 5 represent different settings. The first represents the union of 5 sets, each comprising 3 items, and the second represents the union of 3 sets, each comprising 5 items. The difference between both settings is illustrated below:

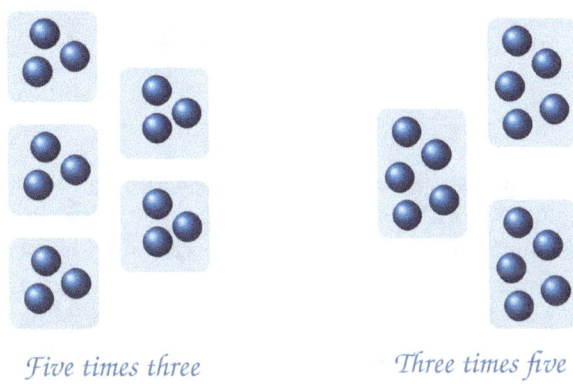

Five times three   *Three times five*

This chapter is only concerned with the multiplication of whole numbers. In this context, interpreting multiplication as repeated addition is most natural. This interpretation must be revised when extending multiplication to fractions, and later on to negative numbers.

7.1.4 Multiplication and counting

Multiplication is closely related to addition. Since addition is closely related to counting, there must also be a close relationship between multiplication and counting.

Think of the meaning of the expression *4 elephants*. It means *an elephant and another elephant and another elephant and another elephant*, or, in short, *4 times elephant*. When we count, we determine how many times a certain object, whatever it is, appears. In repeated addition, the object that we count is itself a set of objects. For example, the expression *3 times 5 sticks* means that the object *5 sticks* appears 3 times. We will see below that this point of view facilitates the interpretation of the properties of multiplication.

7.1.5 Rectangular arrays

We have already seen how helpful good visual representations can be. In this section, we get acquainted with a useful tool for visualizing multiplication: **rectangular arrays**.

The figure below illustrates a rectangular representation of the arithmetic expression 5×4:

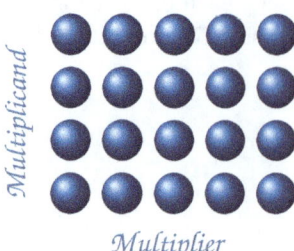

The multiplicand (in this case, 4) is represented by a vertical column of 4 items. This column is copied a number of times indicated by the multiplier (in this case, 5). The 5 copies of a vertical column of 4 items are aligned horizontally and equally spaced, thus creating a rectangular array of items. The total number of items in the array is the value of the product. Below, we will use this visual representation to interpret three properties of multiplication: the commutative property, the associative property and the distributive property.

Please note that since multiplication is commutative, the choice to represent the multiplicand by a vertical column of items rather than by a horizontal row of items is arbitrary.

7.2 Properties of multiplication

Multiplication, just as addition and subtraction, satisfies its own set of rules.

7.2.1 The commutative property

The commutative property of multiplication asserts that the product is not affected by the order of the factors. In order words, if we reverse the multiplier and the multiplicand, the product remains the same. For example:
$$5 \times 7 = 7 \times 5,$$
$$13 \times 462 = 462 \times 13.$$
In algebraic notation, the commutative property asserts that for any two numbers, a and b,
$$a \times b = b \times a.$$

The commutative property of *addition* relies on the very intuitive fact that it does not matter which set we count first (Chapter 4, p. 35). The commutative property of *multiplication* is less intuitive. Examine again the figure on p. 81: is it obvious to you why 3 sets of 5 items contain the same total number of items as 5 sets of 3 items? Isn't is somewhat magical that when you count the total number of items you obtain the same result in both cases?

By means of the rectangular array representation, we can provide an intuitive explanation for the validity of the commutative property of multiplication. Take the arithmetic expression 5×4, for example. The figure below represents this expression by an alignment of 5 vertical columns of 4 items each:

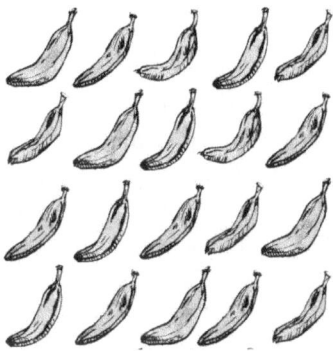

If we rotate this rectangular array, rows turn into columns and columns turn into rows. We obtain a rectangular array of 4 vertical columns of 5 items each. This arrangement represents the arithmetic expression 4×5:

Since rotation does not change the total number of items, it follows that:

$$5 \times 4 = 4 \times 5.$$

Once again, we justified a general property through the examination of a particular example (*proof by a representative example*). This proof is general in the sense that any rectangular array can be rotated regardless of its size.

Please note: The fact that children know that multiplication is commutative in the case of

$$5 \times 4 = 4 \times 5$$

does not necessarily imply that they understand that commutativity holds in any possible case. Most children would say that both sides of the equation are equal because they are both equal to 20. I once challenged a class of third graders with the question as to whether

$$78{,}234 \times 4{,}097 \quad \text{and} \quad 4{,}097 \times 78{,}234$$

were equal. The majority of the children were unable to answer. They lacked the confidence needed to make a statement about a product that they did not know how to evaluate. It takes a great deal of imagination and ability to generalize to come up with an argument stating that there is the same total number of items in a rectangular array that comprises $78{,}234$ columns of $4{,}097$ items each as in a rectangular array that comprises $4{,}097$ columns of $78{,}234$ items each.

7.2.2 The associative property

The associative property asserts that when multiplying three numbers, it does not matter in which order the product is evaluated. In algebraic notation, the associative property asserts that for every three numbers a, b, and c:

$$(a \times b) \times c = a \times (b \times c).$$

For example:

$$\underbrace{\underbrace{(3 \times 4)}_{12} \times 5}_{60} = \underbrace{3 \times \underbrace{(4 \times 5)}_{20}}_{60}.$$

The associative property of multiplication is even less intuitive than the commutative property. There seems to be some magic behind the two calculations yielding the same final result.

Let's examine the above numerical example by means of a word problem:

> There are 3 trucks, 4 crates in each truck, and 5 melons in each crate. How many melons are there in total?

We could approach this question in two different ways:

(a) First, determine the total number of crates: since there are 3 trucks and 4 crates in each truck, then there are 3×4 crates. The total number of melons is given by the total number of crates, times the number of melons in each crate, namely,
$$(3 \times 4) \times 5.$$

(b) First, determine the total number of melons in each truck: since there are 4 crates in each truck with 5 melons in each, then there are 4×5 melons in each truck. The total amount of melons in all trucks combined is 3 times as much, namely,
$$3 \times (4 \times 5).$$

Since both arguments must give the same final answer, we infer that
$$(3 \times 4) \times 5 = 3 \times (4 \times 5).$$

Just as we illustrated the commutative property using a rectangular array, we can use a similar concretization to illustrate the associative property. Consider the three-dimensional structures displayed below. They are identical, up to a different coloring.

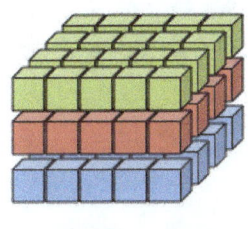

The coloring of the top structure exhibits three layers, each being a 4-by-5 array of cubes. Thus, the total number of cubes in the top structure is
$$3 \times (4 \times 5).$$

The coloring of the bottom structure exhibits a 3-by-4 array of rods, each comprising 5 cubes. Thus, the total number of cubes in the bottom structure is
$$(3 \times 4) \times 5.$$

Since the number of cubes in both structures is the same, we conclude that
$$3 \times (4 \times 5) = (3 \times 4) \times 5.$$

7.2.3 The distributive property

The distributive property connects addition and multiplication. It asserts that for any three numbers a, b, and c:
$$(a + b) \times c = a \times c + b \times c.$$
For example:
$$\underbrace{\underbrace{(5 + 3)}_{8} \times 4}_{32} = \underbrace{\underbrace{5 \times 4}_{20} + \underbrace{3 \times 4}_{12}}_{32}.$$

The distributive property stems directly from the interpretation of multiplication as representing the repeated occurrence of equally-sized sets. We can interpret the above numerical example as follows: 5 objects plus 3 objects make 8 objects, as long as *object* refers in both cases to entities of the same type. Now let us replace the word *object* by *4 apples*. We obtain that *5 times 4 apples* plus *3 times 4 apples* make *8 times 4 apples*.

The distributive property can also be visualized using rectangular arrays. The figure below illustrates the identity $(3 + 5) \times 4 = 5 \times 4 + 3 \times 4$:

$$5 \times 4 \qquad 3 \times 4 \qquad (5 + 3) \times 4$$

The distributive property holds also with addition replaced by subtraction. For any three numbers, a, b, and c:
$$(a - b) \times c = a \times c - b \times c.$$
For example:
$$\underbrace{\underbrace{(12 - 3)}_{9} \times 8}_{72} = \underbrace{\underbrace{12 \times 8}_{96} - \underbrace{3 \times 8}_{24}}_{72}.$$

The identity between both sides can be interpreted as follows: 12 objects minus 3 objects make 9 objects, as long as *object* refers in both cases to entities of the same type. Now let us replace the word *object* object by *8 apples*. We obtain that *12 times 4 apples* minus *3 times 8 apples* make *9 times 8 apples*.

Since multiplication is commutative, the distributive law can be formulated also in the following ways:
$$a \times (b + c) = a \times b + a \times c$$
$$a \times (b - c) = a \times b - a \times c.$$

7.2.4 Laws of variation

Another property of multiplication is that if we vary either the multiplier or the multiplicand by a multiplicative factor, then the product varies by that same multiplicative factor. For example:

If $4 \times 6 = 24$ then $(2 \times 4) \times 6 = 2 \times 24$.

And, similarly,

If $4 \times 6 = 24$ then $4 \times (6 \times 3) = 24 \times 3$.

These are actually reiterations of the associative property. Rather than interpreting these identities as properties of products of three numbers, we interpret them as rules for how the product of two factors changes when one of the factors varies multiplicatively. This slightly different point of view is often more intuitive, and furthermore, turns out to be useful when considering together both operations of multiplication and division (see the next chapter).

7.2.5 The identity property of one

One times 5 is 5, and 5 times 1 is 5. The number 1 has the following property: For any number a:

$$a \times 1 = 1 \times a = a.$$

The number 1 is the **identity element** of multiplication: when 1 is a factor in a product, the product is equal to the other factor. Please note the analogy between 0 as the identity element of addition and 1 as the identity element of multiplication.

7.2.6 Multiplication by zero

How much is 5 times 0? Since we may interpret multiplication as repeated addition:

$$5 \times 0 = 0 + 0 + 0 + 0 + 0.$$

And so the answer is 0. It is clear that the product would have been 0 if the multiplier were any other number.

What happens when the multiplier is 0? How much is, for example, 0 times 5? In other words: How many items are there in 0 sets comprising 5 items each? The answer is 0, regardless of the multiplicand.

Thus, 0 has the property that for any number a:

$$a \times 0 = 0 \times a = 0.$$

Please note that we already knew that multiplication was commutative. Therefore, if a zero multiplicand implies a zero product, then it must also be the case that a zero multiplier implies a zero product. Yet, properties of operations are too abstract to serve as convincing arguments at an early age. For children to truly understand multiplication by zero, they must possess concrete mental representations of both $a \times 0$ and $0 \times a$.

7.3 Word problems

7.3.1 Repeated addition

Most word problems that involve multiplication have a similar structure: the multiplicand represents some quantity and the multiplier represents the number of times that that quantity appears. For example:

> There are 8 pencils in a pack. How many pencils are there in 6 packs?

Multiplication word problems are a good opportunity to combine arithmetic with other subjects, such as measurements. For example:

> The running track is 400 meters long. What distance did I run if I completed 5 laps?

Or:

> There are 60 seconds in a minute, and 60 minutes in an hour. The train is scheduled to arrive in 2 hours. How many minutes will elapse until the train's arrival? How many seconds?

7.3.2 Combinatorial problems

Combinatorics is a branch of mathematics concerned with counting the number of ways certain sets can be arranged. Combinatorial problems occur in many situations, and the solution of many of them is based on multiplication. The study of multiplication is therefore a natural opportunity to get acquainted with combinatorial problems.

Let's start with something really simple. Consider the following question:

> There are 3 hats: one red, one green and one blue. How many different possibilities are there to pick one hat out the three?

The answer is, of course, 3, since we can pick either the red hat, or the green hat, or the blue hat.

Now let us move to a higher level of complexity:

> In addition to the 3 hats, there are 4 shirts: red, green, blue and yellow (see figure below). How many different possibilities are there to pick a pair of garments consisting of one hat and one shirt?

We can start counting all the possible combinations, but this is not a practical strategy with larger numbers. It also becomes difficult to keep track of which combinations we already counted and which we didn't. A systematic way of counting the number of combinations is to use a table like the following:

Hat/Shirt	Red	Green	Blue	Yellow
Red				
Green				
Blue				

Each row in the table represents a hat and each column in the table represents a shirt, so that each entry in the table represents a combination of a hat and a shirt. This representation reveals that the number of possible combinations equals the number of entries in the table, which is the product of the number of hats and the number of shirts, i.e., $3 \times 4 = 12$.

An alternative way for representing all possible combinations is by a ***tree diagram***:

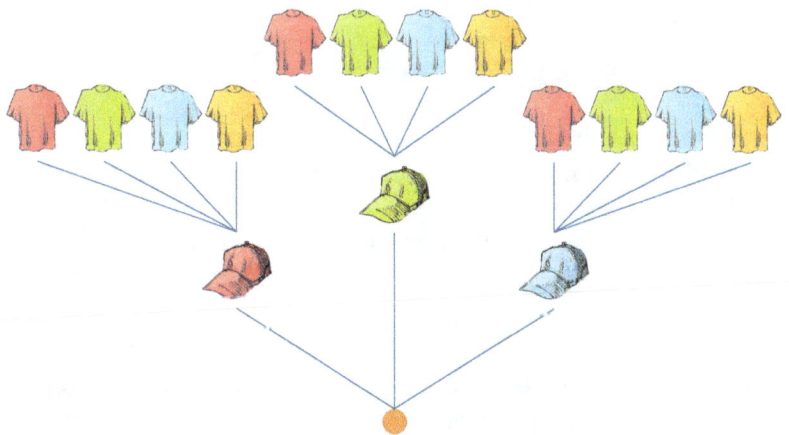

How do we read this tree? Imagine a decision-making procedure. At the start, we are located at the **root** of the tree, which is the dot at the bottom of the diagram. We first select a hat. There are 3 different possibilities, each represented by a **branch** that ascends from the root to one of the 3 hats. Once we have selected a hat, we move up the tree according to our choice. We then select a shirt. There are 4 different possibilities, each represented by a branch that ascends to one of the 4 shirts. Once we have selected a shirt, we move up the tree according to our choice. We have thus reached a **leaf**, which is a location from which no more branches ascend. Each leaf in this tree has a trajectory that connects it to the root of the tree, and this trajectory corresponds to a unique choice of a hat and a shirt. Therefore, the number of possible combinations equals the number of leaves, which is 12.

Please note that from a mathematical point of view, it is totally unimportant what the colors are. We could have as well named them color #1, color #2, etc. In fact, it doesn't even matter that the attribute we attach to each garment is a color, nor that we are dealing with combinations of garments.

Now let us add another layer to the problem:

> In addition to 3 hats and 4 shirts, there are also 5 pair of pants (see figure below). How many different possibilities are there to pick three garments consisting of one hat, one shirt and one pair of pants?

There are several possible ways to solve this problem. One way is to represent all the possible combinations in a three-dimensional table of 5 layers, one layer for each pair of pants. Each of the 5 layers will be a two-dimensional table comprising 4 lines and 3 rows. In this way, we can see that the number of possible combinations is:

$$3 \times 4 \times 5 = 60.$$

Another way is to count on the previous knowledge that there are 12 possible hat/shirt combinations, and to treat the problem as requiring the selection of only two items: One bundled item (hat and shirt), of which there are 12, and pair of pants, of which there are 5. Since we already know that the number of combinations is the product of both quantities, then the total number of combinations in this case is:

$$12 \times 5 = 60.$$

Please note: Combinatorial problems are presented here as a class of word problems whose solutions involve multiplication. In these problems, unlike the more common multiplication problems, both factors have equal status, that is, there is no natural distinction between a multiplier and a multiplicand. Another class of problems in which both factors have equal status is the calculation of areas (see Chapter 18).

7.4 Evaluating multiplication

7.4.1 If you can count, you can multiply

Just as for addition and subtraction, here too we claim: If you can count, you can multiply.

The most straightforward way to turn a multiplication problem into a counting task, is based, once again, on rectangular arrays. Consider, for example, the arithmetic expression 4×3. It can be represented as a rectangular array of 4 columns comprising 3 elements each:

Now all we have to do is count the symbols.

Another way to visualize the same product is to draw 3 horizontal lines, intersect them with 4 vertical lines, and count the points of intersection:

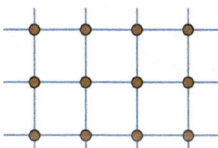

Such methods of evaluation may not be efficient, but they can always serve as a last resort.

7.4.2 The multiplication table

The evaluation of products is significantly harder than the evaluation of sums. In fact, until a few centuries ago, the evaluation of products of multi-digit numbers was considered too difficult for the average adult. Today, in many countries the evaluation of such products is part of third and fourth grade curricula. In lower grades, children are expected to develop competency in calculating products of single-digit numbers, and there is for a good reason for this: the evaluation of products of multi-digit numbers is based on repeated evaluations of products of single-digit numbers. The collection of all products of single-digit numbers can be organized in a table known as the **multiplication table** (cf. addition table, Chapter 4, p. 45):

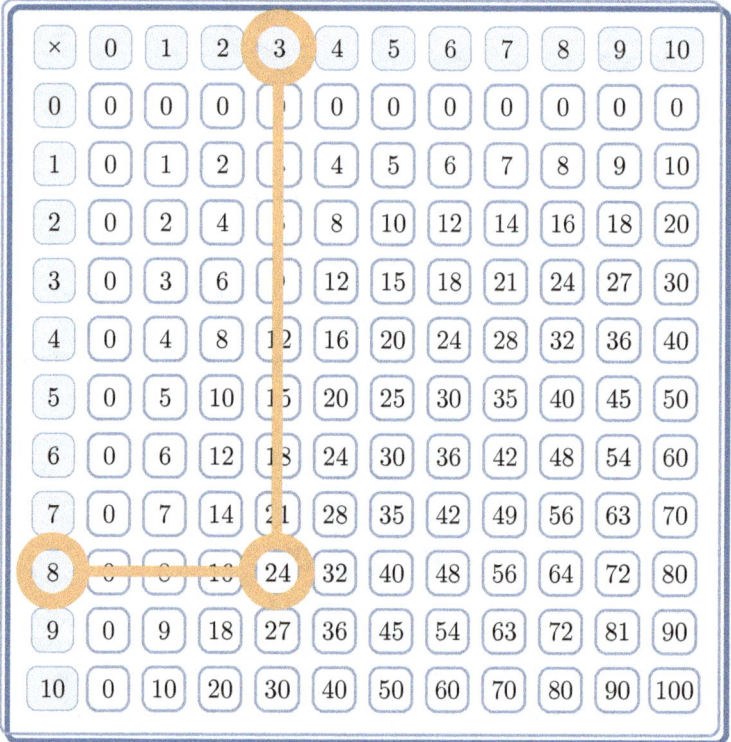

For example, to evaluate the product 3×8 one has to draw a vertical line from the column headed by the number 3, and a horizontal line from the row headed by the number 8. The meeting point of these two lines indicates the product—in this case, 24.

7.4.3 Various evaluation strategies

The memorization of the multiplication table is a fairly easy task for some children, but a very difficult and frustrating task for others. Moreover, learning a set of multiplication facts by heart does not necessarily indicate that a child understands the interpretations and the properties of multiplication. The early learning stages of multiplication offer opportunities to develop a myriad of evaluation strategies. These serve both to consolidate the child's understanding of multiplication, and as a step toward fast and effective evaluation skills. In this section, we present several evaluation strategies. It is very likely that children will come up with many others.

7.4.3.1 Skip counting

The most basic way of calculating products is by skip counting. Suppose we want to evaluate the product:
$$6 \times 7,$$
which corresponds to the repeated addition:
$$7 + 7 + 7 + 7 + 7 + 7.$$
If we were to demonstrate this repeated addition on the number line, we would start at 0, and perform 6 leaps of 7 points at a time:

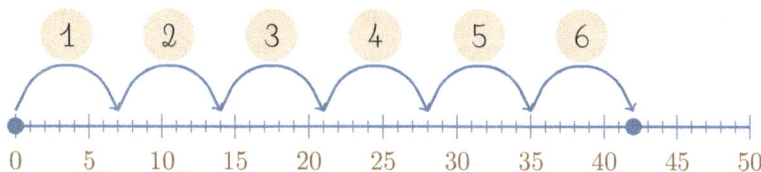

Skip counting yields the following sequence of multiples of 7:
$$7, 14, 21, 28, 35, 42.$$
Thus, the product equals 42.

7.4.3.2 Using the laws of variation

Suppose that we already know that
$$3 \times 9 = 27.$$
Can we exploit this knowledge to evaluate how much is
$$6 \times 9?$$
Yes. It follows from the laws of variation that:
$$6 \times 9 = (2 \times 3) \times 9 = 2 \times (3 \times 9) = 2 \times 27 = 54.$$
Some children will probably reason as follows: if 3 times 9 is 27, then 6 times 9 is twice as much.

7.4.3.3 Using the distributive property

Suppose, for example, that a child is proficient in the evaluation of products with both factors up to 5. Then, products with larger factors can be evaluated using the distributive property. For example,

$$7 \times 5 = (4+3) \times 5 = 4 \times 5 + 3 \times 5 = 20 + 15 = 35.$$

Another common use of the distributive property is for calculating multiples of 9, for example,

$$9 \times 7 = (10-1) \times 7 = 10 \times 7 - 1 \times 7 = 70 - 7 = 63.$$

Or, in words: If 10 times 7 is 70, then 9 times 7 is 7 fewer than 70.

7.4.3.4 Additional Strategies

Children often adopt a variety of "tricks" that help memorize products: For example:

- **Multiples of 9**: One of the characteristics of multiples of 9 (up to 9×9), is that the sum of both digits of the product is 9, and the tens digit of the product is 1 less than the other factor. For example, in 7×9, the tens digit of the product is 6 (1 less than 7), and the ones digit sums up to 9 together with the tens digit, hence it is 3, that is, $7 \times 9 = 63$.

- **Multiples of 5**: Multiples of 5 have the property that the ones digit is 5 if the other factor is odd and 0 if the other factor is even. Knowing this fact does not provide a systematic way for evaluating multiples of 5, but it reduces significantly the range of plausible solutions, and as such provides a useful validation tool.

7.4.4 Multiplication by 10

Multiplication by 10 is the most important operation in preparation for the multiplication of multi-digit numbers. The rule we teach at school is as follows: *To multiply a number by 10, just append a zero to its right.* For example:

$$423 \times 10 = 4230.$$

Where does this rule stem from? Consider the expression:

$$6 \times 10.$$

We can view it literally as 6 times 10, but we can also view is as 6 tens. Our decimal representation of 6 tens places the digit 6 in the tens column and the digit 0 as a place holder in the ones column. Notice that we have thus obtained that $6 \times 10 = 60$ without having really calculated anything.

Consider next the expression:

$$12 \times 10.$$

By the distributive property, 12 tens are 10 tens plus 2 tens. 10 tens are a hundred, so this expression equals 1 hundred plus 2 tens, whose decimal representation is 120.

More such applications of the distributive property show the product of any number by 10 is obtained by shifting every digit of that number one place to the left, and writing 0 in the ones column.

Mathematical problems

Problem 7.1. Consider the following list of equations:

$$2 \times 2 = 4 \qquad 1 \times 3 = 3$$
$$3 \times 3 = 9 \qquad 2 \times 4 = 8$$
$$4 \times 4 = 16 \qquad 3 \times 5 = 15$$
$$5 \times 5 = 25 \qquad 4 \times 6 = 24$$
$$6 \times 6 = 36 \qquad 5 \times 7 = 35.$$

(a) What rule is apparent in this list?
(b) Deduce how much is 12×14 given that $13 \times 13 = 169$.
(c) Explain why this rule holds.

Problem 7.2. Draw a tree diagram for choosing a hat, a shirt and a pair of pants, when your wardrobe consists of 3 hats, 4 shirts and 2 pairs of pants. Explain how one can infer from this tree how many possible combinations there are.

Problem 7.3. In how many different ways can two children stand in line? Three children? Four? Ten?

Pedagogical problems

Problem 7.4. Teacher Kate asked her students the following question:

> In the grove there are 18 rows of trees. Each row comprises 21 apple trees and 17 lemon trees. How many trees are there in the grove?

Teacher Kate asked each child to write an appropriate arithmetic expression.

Gili wrote: $18 \times (21 + 17)$.
Alon wrote: $18 \times 21 + 18 \times 17$.

(a) Are both expressions appropriate?
(b) What does the expression 21 + 17 in Gili's solution represent?
(c) What does the expression 18 × 21 in Alon's solution represent?

Problem 7.5. A common mistake children make is displayed in the following (erroneous) evaluation:
$$9 \times 8 = 10 \times 8 - 9 = 80 - 9 = 71.$$

Find the origin of the mistake, and propose a way to resolve it.

Chapter 8

Division

Division is the fourth and last arithmetic operation taught in school, and it is considered the most challenging. The key to overcoming children's difficulties is to provide them with a profound understanding of the operation. This chapter emphasizes the different interpretations of division and the properties that ensue from these interpretations. A successful assimilation of division also plants the seeds for the learning of fractions in the near future.

8.1 What is division?

Just as subtraction is the inverse operation of addition, division is the inverse operation of multiplication. Like subtraction, division has several interpretations, the most frequent of which are known as **partitive division** and **quotative division**. Both interpretations can be viewed as multiplication problems in which one of the factors is unknown. Throughout this chapter, I encourage you to look for the analogy between subtraction as an unknown-addend problem and division as an unknown-factor problem.

8.1.1 Division as a model for sharing

Consider the following word problem:

> 18 carrots were divided equally among 3 bunnies. How many carrots did each bunny get?

In this problem, there is a given number of objects (18 carrots), shared equally into a given number of sets (3, as the number of bunnies). We have to find out the number of objects in each set (i.e., the size of each **share**, or **portion**, or **quota**).

The fact that the sets are of equal size relates this division problem to multiplication. In fact, this word problem can be represented by the following arithmetic equation:

$$\underbrace{3}_{\text{number of bunnies}} \times \underbrace{\boxed{?}}_{\text{carrots per bunny}} = \underbrace{18}_{\text{total number of carrots}}$$

This is a multiplication equation in which one of the factors is not known. Here, the unknown factor is the **multiplicand**, which represents the size of each share. Division in which a set of given size is divided equally into a known number of shares is called **partitive division**.

8.1.2 Division as a model for rationing

Consider the following word problem:

> 30 children gathered for a basketball tournament. The children were divided into teams of 5 players each. How many teams participated in the tournament?

In this problem, there is a given number of objects (30 children), partitioned into equal sets. We know the number of objects in each set (5 children), that is, we are given the size of each quota (or portion, or share). This time we have to find out the number of sets. This word problem can be represented by the following arithmetic equation:

$$\underbrace{\boxed{?}}_{\text{number of teams}} \times \underbrace{5}_{\text{children per team}} = \underbrace{30}_{\text{total number of children}}$$

Once again, we face a multiplication equation in which one of the factors is not known. This time the unknown factor is the **multiplier**, which represents the number of sets. Division in which a set of given size is divided into quotas of a known size is called **quotative division**.

8.1.3 The division operation

Now that we have seen two representative situations of division, we can formally define the operation.

The division sign is ÷ (**divided by**). The first operand is called the **dividend** (a quantity being divided) and the second operand is called the **divisor**. The result of the operation is called the **quotient**.

Consider, for example, the arithmetic expression:

$$18 \div 3.$$

Its value is defined as the unknown factor that solves either of the following equation:

$$3 \times \boxed{?} = 18 \quad \text{or} \quad \boxed{?} \times 3 = 18.$$

Thus,

$$18 \div 3 = \boxed{6}$$

because

$$3 \times \boxed{6} = 18 \quad \text{or because} \quad \boxed{6} \times 3 = 18.$$

In algebraic notation, division is defined as follows: For two numbers which we denote by a and b, the algebraic expression:

$$a \div b$$

is equal to the unknown that solves either of the unknown-factor equations:

$$b \times \boxed{?} = a \quad \text{or} \quad \boxed{?} \times b = a.$$

Please note: At this point, when we write $a \div b$ we assume that there exists a whole number that solves the unknown-factor equation: $\boxed{?} \times b = a$. Division with a remainder and division resulting in a fraction will be presented in Volume 2.

8.2 Word problems

8.2.1 Modeling with division

Division has several interpretations that are inherently different. Finding the size of a share (partitive division) and finding the number of shares (quotative division) are very different tasks. It is actually quite confusing that both tasks are represented by the same arithmetic operation.

This is a good moment to recapitulate the two interpretations of division, with a representative word problem for each interpretation:

(a) **Partitive division**: 56 pancakes are partitioned evenly onto 7 plates. How many pancakes are on each plate?
(b) **Quotative division**: 56 pancakes are partitioned evenly, 7 pancakes on each plates. How many plates are there?

The solutions to both word problems are represented by the same arithmetic expression:

$$56 \div 7.$$

In partitive division, we are given the number of shares into which a total amount was partitioned, and we have to find out the size of each share. An illustration of this is given below:

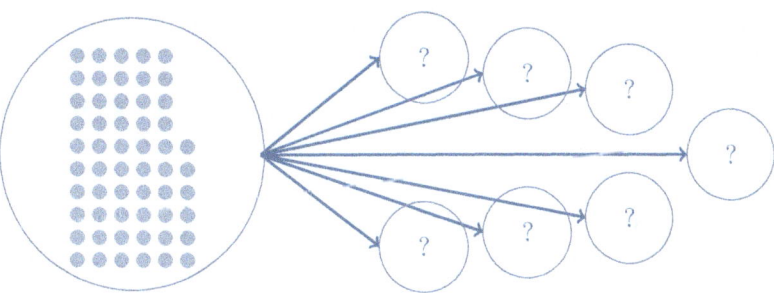

In quotative division, we are given the size of each share (or quota), and we have to find how many such shares there are. An illustration of this is given below:

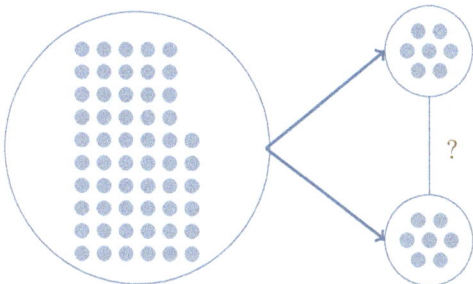

8.2.2 Interpretation

We have already seen the advantages of multiple interpretations being attributed to an arithmetic operation: given an evaluation problem, we have the freedom to choose a word problem that best guides our intuition.

To demonstrate how this advantage can be exploited for division, consider the following two arithmetic expressions:

$$18 \div 2 \quad \text{and} \quad 18 \div 9.$$

It may seem natural to represent both expressions using the same metaphor since both involve the same operation.

Consider first the arithmetic expression $18 \div 2$. Its value can be interpreted as the solution to either of the following two questions:

(1) **Partitive division**: 18 objects are partitioned evenly into 2 sets. How many objects are in each set?
(2) **Quotative division**: 18 objects are partitioned evenly, 2 objects in each set. How many sets are there?

It turns out that most of us find it easier to visualize division problems that involve a smaller number of sets. Hence, most of us will choose to represent this problem using partitive division.

Next, consider the arithmetic expression $18 \div 9$. Its value can be interpreted as the solution to either of the following two questions:

(1) **Partitive division**: 18 objects are partitioned evenly into 9 sets. How many objects are in each set?
(2) **Quotative division**: 18 objects are partitioned evenly, 9 objects in each set. How many sets are there?

This time the number of sets is smaller if we interpret the problem as quotative division. Indeed, this is the interpretation most of us will choose.

With that in mind, I recommend exposing children to both interpretations of division, making sure that they develop the ability to interpret abstract division problems in either way. The richer their toolbox is, the better they will understand and the better they will perform.

8.2.3 Sneak preview: Fraction division

Most people tend to interpret division only in terms of partitive division. Being restricted to a single interpretation is a serious limitation when it comes to dividing fractions (a subject we study in depth in Volume 2). For example, consider the arithmetic expression:

$$\frac{1}{2} \div \frac{1}{4}.$$

An attempt to interpret it through partitive division may look as follows:

> Half an apple was equally partitioned into a quarter of a set. How many apples are in each set?

This question sounds absurd (even though, as we will see in Volume 2, it is possible to interpret this arithmetic expression using partitive division).

An attempt to interpret the same expression through quotative division may look as follows:

> Half an apple was partitioned into quotas of a quarter of an apple each. How many such quotas are there?

Or

> How many quarters of an apple are contained in a half of an apple?

This time, the question is meaningful, and the answer is not hard to figure out: 2.

 **Activity: Interpretations of division**

Ask a group of children to evaluate the arithmetic expression $40 \div 5$ using material objects, such as sticks.

(a) Ask one group of children to perform partitive division: partition the sticks evenly into 5 sets, and count how many sticks are in each set.
(b) Ask another group of children to perform quotative division: partition the sticks evenly into sets of 5 sticks each, and count the number of sets.
(c) Observe how, despite the difference between the tasks, the same numerical result is obtained.
(d) Discuss whether this is always so: repeat both procedure with 42 sticks, partitioned first into 6 sets, and then into quotas of 6 sticks.
(e) Relate the equivalence of both practices with the commutative property of multiplication.
(f) Notice how similar this activity is to the activity proposed on p. 72 in the context of even and odd numbers.

8.3 Properties of division

Division bears a strong similarity to subtraction as both operations are inverses of another operation: Subtraction is the inverse of addition, and division is the inverse of multiplication. The similarity between these two operations is reflected in their respective properties. I encourage you to compare this section with the section *The properties of subtraction* in Chapter 5.

8.3.1 Division is not commutative

When we interchange the dividend and the divisor, the result (i.e., the quotient) changes. In algebraic notation:
$$a \div b \neq b \div a.$$
In other words, division does not satisfy the commutative property, or in short, division is not commutative.

8.3.2 Division is not associative

When we evaluate two consecutive division operations, the order of evaluation matters. In other words, the result depends on which operation is evaluated first. Consider, for

example, the following arithmetic expression:
$$80 \div 8 \div 2.$$
If we first divide 80 by 8 and then divide the resulting quotient by 2, we obtain:
$$\underbrace{(80 \div 8)}_{10} \div 2 = 5.$$
In contrast, if we first divide 8 by 2 and then divide 80 by the resulting quotient, we obtain:
$$80 \div \underbrace{(8 \div 2)}_{4} = 20.$$
Both ways of evaluation yield different results.

The convention here is the same as for subtraction: in the absence of parentheses, the operations should be performed in the order in which they appear, from left to right. In the above example,
$$80 \div 8 \div 2 = 5.$$

8.3.3 The distributive property

We have already encountered the distributive property in Chapter 7, as a property that links multiplication and addition. A distributive property also links division and addition.

8.3.3.1 Distributivity of the dividend

Division satisfies a distributive property with respect to the dividend: for every three numbers, a, b and c:
$$(a + b) \div c = a \div c + b \div c.$$
For example:
$$\underbrace{\underbrace{(18 + 6)}_{24} \div 3}_{8} = \underbrace{\underbrace{18 \div 3}_{6} + \underbrace{6 \div 3}_{2}}_{8}.$$

The distributive property becomes intuitive in concrete situations. For example:

> 18 apples and 6 oranges were distributed evenly between 3 children. How many pieces of fruit did every child get?

On the one hand, if a total of $18 + 6$ pieces of fruit were distributed evenly between 3 children, then every child got
$$(18 + 6) \div 3 \quad \text{fruits.}$$

On the other hand, if 18 apples were distributed evenly between 3 children, and then 6 oranges were distributed evenly between those 3 children, then every child got
$$18 \div 3 \quad \text{apples} \; + \; 6 \div 3 \quad \text{oranges}.$$
Since both ways of reasoning must yield the same answer, we conclude that
$$(18 + 6) \div 3 = 18 \div 3 + 6 \div 3.$$

Similar to multiplication, the distributive property holds also if addition is replaced by subtraction. For every three numbers, a, b and c:
$$(a - b) \div c = a \div c - b \div c.$$
For example:
$$(\underbrace{18 - 6}_{12}) \div 3 = \underbrace{18 \div 3}_{6} - \underbrace{6 \div 3}_{2}.$$

8.3.3.2 Sneak preview: Division as multiplication by a fraction

Fifth graders in the United States learn that division is equivalent to multiplication by a fraction. For example, division by 3 is equivalent to multiplication by a third. Therefore, the identity:
$$(18 + 6) \div 3 = 18 \div 3 + 6 \div 3$$
can also be written as:
$$\frac{1}{3} \times (18 + 6) = \frac{1}{3} \times 18 + \frac{1}{3} \times 6.$$
Viewed this way, the distributivity of the dividend is an instance of the distributive property of multiplication.

8.3.3.3 Non-distributivity of the divisor

The distributive property of multiplication holds whether we distribute the multiplier or the multiplicand. That is,
$$(a + b) \times c = a \times c + b \times c,$$
and also
$$a \times (b + c) = a \times b + a \times c.$$
Clearly, if one property holds then the other must also hold, since multiplication is commutative.

Division is not commutative, and therefore the fact that the dividend satisfies a distributive property does not necessarily imply that the divisor does too. In fact, the divisor is not distributive, namely:
$$a \div (b + c) \neq a \div b + a \div c.$$
For example:
$$24 \div \underbrace{(4 + 2)}_{6} \neq \underbrace{24 \div 4}_{6} + \underbrace{24 \div 2}_{12}.$$

8.3.3.4 Sneak preview: Distributivity in fractions

When performing operations on fractions, and when later manipulating algebraic fractions, understanding the distributive law of division is essential for understanding why

$$\frac{24 + 96}{6} = \frac{24}{6} + \frac{96}{6},$$

whereas

$$\frac{144}{6 + 12} \neq \frac{144}{6} + \frac{144}{12}.$$

8.3.4 Laws of variation

In this section, we examine how multiplicative variations in both dividend and divisor affect the quotient. Together with the laws of variation described in Chapter 7, this combined set of laws forms the basis of what is known as **proportional reasoning**. Proportionality between two quantities is synonymous to one quantity being the product of the other quantity by a fixed factor. For example, the price of gasoline in dollars is proportional to the quantity of gasoline in gallons. Proportional reasoning, a term coined by Swiss developmental psychologist Jean Piaget (1896–1980), refers to the understanding of how one quantity changes as the result of another quantity changing, given that they are proportional to each other. Proportional reasoning is a very important concept that comes into play in many applications, such as in physics and economics.

We describe below four laws of variation, which all arise from examining a quotient, $a \div b$, and asking how it changes when a and b are either multiplied or divided by a given factor.

8.3.4.1 Multiplication of the dividend

Multiplication of the dividend results in the multiplication of the quotient by the same factor.

In algebraic terms, if a was originally the dividend and b was originally the divisor, and we multiplied the dividend by c, then the new quotient is equal to the original quotient multiplied by c. In algebraic notation:

$$(a \times c) \div b = (a \div b) \times c.$$

For example:

$$\underbrace{(18 \times 3)}_{54} \div 2 = \underbrace{(18 \div 2)}_{9} \times 3.$$
$$ 27 27$$

This fact is easily interpreted in terms of partitive division: if we divide 3 times as much into the same number of shares, each share is 3 times larger.

8.3.4.2 Multiplication of the divisor

Multiplication of the divisor results in the division of the quotient by the same factor.

In algebraic terms, if a was originally the dividend and b was originally the divisor, and we multiplied the divisor by c, then the new quotient is equal to the original quotient divided by c:

$$a \div (b \times c) = (a \div b) \div c.$$

For example:

$$72 \div \underbrace{(4 \times 3)}_{12} = \underbrace{(72 \div 4)}_{18} \div 3.$$

$$\underbrace{}_{6} = \underbrace{}_{6}$$

The interpretation is as follows: if we divide the same amount into 3 times as many shares, each share is 3 times smaller.

8.3.4.3 Division of the dividend

Division of the dividend results in the division of the quotient by the same factor.

In algebraic terms, if a was originally the dividend and b was originally the divisor, and we divided the dividend by c, then the new quotient is equal to the original quotient divided by c:

$$(a \div c) \div b = (a \div b) \div c.$$

For example:

$$\underbrace{(48 \div 3)}_{16} \div 8 = \underbrace{(48 \div 8)}_{6} \div 3.$$

$$\underbrace{}_{2} = \underbrace{}_{2}$$

In simple terms, if the amount that we divide is smaller by a factor of 3, then so is the size of each share.

8.3.4.4 Division of the divisor

Division of the divisor results in the multiplication of the quotient by the same factor.

In algebraic terms, if a was originally the dividend and b was originally the divisor, and we divided the divisor by c, then the new quotient is equal to the original quotient multiplied by c:

$$a \div (b \div c) = (a \div b) \times c.$$

For example:

$$48 \div \underbrace{(6 \div 3)}_{2} = \underbrace{(48 \div 6)}_{8} \times 3.$$

$$\underbrace{}_{24} = \underbrace{}_{24}$$

In simple terms, if there are 3 times fewer parts, then the size of each portion increases by a factor of 3.

The analogy between the addition-subtraction relation and the multiplication-division relation, as manifested in the laws variation, is summarized in the following table:

	Add/Sub	Mult/Div	
Increase minuend	$(a+c)-b=(a-b)+c$	$(a\times c)\div b=(a\div b)\times c$	Multiply dividend
Increase subtrahend	$a-(b+c)=(a-b)-c$	$a\div(b\times c)=(a\div b)\div c$	Multiply divisor
Decrease minuend	$(a-c)-b=(a-b)-c$	$(a\div c)\div b=(a\div b)\div c$	Divide dividend
Decrease subtrahend	$a-(b-c)=(a-b)+c$	$a\div(b\div c)=(a\div b)\times c$	Divide divisor

8.3.5 Interchanging multiplication and division

In Chapter 5, we saw that interchanging the order of the operands in a chain of addition and subtraction operations does not affect the end result, as long as each operand retains the operation that preceded it. A similar rule holds for chains of multiplication and division operations.

For example:

$$90 \times 3 \div 5 \div 6 \times 2 = 90 \div 6 \times 2 \div 5 \times 3.$$

With the left side: $90 \times 3 = 270$, $270 \div 5 = 54$, $54 \div 6 = 9$, $9 \times 2 = 18$.

With the right side: $90 \div 6 = 15$, $15 \times 2 = 30$, $30 \div 5 = 6$, $6 \times 3 = 18$.

8.3.6 Adjoint division equation

Suppose we are given the following identity:

$$54 \div 6 = 9.$$

Then, the following identity also holds:

$$54 \div 9 = 6.$$

The second equation was derived from the first by transposing the divisor and the quotient. We say that the second equation is the **adjoint** of the first, or that both equations are adjoint to each other.

The first equation implies the second equation because both derive from the same multiplication equation:

$$6 \times 9 = 54.$$

Differently stated, the value of

$$54 \div 6$$

is the number that when multiplied by 6 yields 54. If we know that number, 9, then we automatically know the number that when multiplied by 9 yields 54. It is 6.

8.3.7 Division of zero and division by zero

Most people know that one cannot divide by zero. (It is one of those mantras that you learn at school.) Others wonder whether zero can be divided, or whether it makes any sense to divide 0 by 0. This section is devoted to properties of zero in the context of division.

8.3.7.1 Division of zero

Consider the following arithmetic expression:

$$0 \div 4.$$

Is this a valid expression? Does it have a numerical value?

To answer this question, we can try to interpret it from the perspective of partitive division:

> 0 pancakes are divided evenly among 4 children. How many pancakes does each child get?

Or from the perspective of quotative division:

> 0 pancakes are partitioned into sets of 4 pancakes each. How many sets are there?

In both cases, the obvious answer is zero.

The use of concrete examples to interpret a mathematical expression is a good habit. Yet, one has to be careful, as it sometimes happens that gaps between the formal mathematical language and the casual spoken language lead us astray. In order to make sure that our reasoning is correct, we have to return to the mathematical definition of division: the value of the arithmetic expression $0 \div 4$ equals the unknown in the unknown-factor equation

$$4 \times \boxed{?} = 0.$$

Since 0 is the unique number that when multiplied by 4 yields 0, we conclude that $0 \div 4 = 0$.

8.3.7.2 Division by zero

Consider the following arithmetic expression:

$$4 \div 0.$$

Is this a valid expression? Does it have a numerical value?

To answer this question, we can try to interpret it from the perspective of partitive division:

> 4 pancakes are divided evenly among 0 children. How many pancakes does each child get?

Or from the perspective of quotative division:

> 4 pancakes are partitioned into sets of 0 pancakes each. How many sets are there?

Both questions sound absurd. What is the conclusion? That the expression $4 \div 0$ is meaningless? We should refrain from hasty conclusions. After all, operations involving zero have always challenged people's intuition.

What about the following (erroneous) solution?

> If 4 pancakes are divided evenly among 0 children, then nobody gets any pancake, and therefore $4 \div 0 = 0$.

Or another (erroneous) interpretation:

If 4 pancakes are divided evenly among 0 children, no one gets any pancake, and all 4 pancakes remain. Therefore 4 ÷ 0 = 4.

The mathematical approach to resolve confusion is to turn back to definitions. By definition, the value of the arithmetic expression 4 ÷ 0 equals the unknown in the unknown-factor equation

$$0 \times \boxed{?} = 4.$$

This unknown is a number, which, when multiplied by 0, yields 4. Such a number does not exist, since 0 times any number is 0. Hence, we conclude that the arithmetic expression 4 ÷ 0 cannot be assigned any numerical value. We say that it is **undefined**. (People often say that division by zero is *not allowed*. Nobody can forbid you from writing an arithmetic expression with zero as divisor; it is not punishable by law... you just obtain an expression that has no numerical value.)

8.3.7.3 Division of zero by zero

When we divide zero by a number, we get zero. When we divide by zero, we get an undefined expression. We now confront these two rules, and ask ourselves whether it makes sense to divide 0 by 0. Do we get zero? Do we get an undefined expression?

Let's interpret the arithmetic expression 0 ÷ 0 from the perspective of partitive division:

> 0 pancakes are divided evenly between 0 children. How many pancakes does each child get?

Isn't the answer zero?

As in the previous two cases, we resolve the question by turning back to the definition of division. The value of the arithmetic expression 0 ÷ 0 equals the unknown in the unknown-factor equation

$$0 \times \boxed{?} = 0.$$

This unknown is a number that when multiplied by 0 yields 0. This time we face a new situation where *every number* satisfies this property. Thus, it may seems as if any number could be equal to $0 \div 0$. However, since every arithmetic expression must be uniquely defined, the convention is that the quotient $0 \div 0$ is undefined as well.

To summarize: Division by 0 is not defined, whether the dividend is zero or not.

8.3.8 Division by 1

Consider, for example, the arithmetic expression $17 \div 1$. Its value can be interpreted as the solution to either of the two questions:

(1) **Partitive division**: 17 objects are partitioned equally between one set (in fact, they are not partitioned at all). How many objects are in each set?
(2) **Quotative division**: 17 objects are partitioned into sets, each containing one object. How many sets are there?

In either case, the answer is obviously 17.

When the divisor is 1, the quotient is always equal to the dividend. In algebraic notation, for any number a:

$$a \div 1 = a.$$

8.3.9 Division of a number by itself

Consider, for example, the arithmetic expression $17 \div 17$. Its value can be interpreted as the solution to either of the two questions:

(1) **Partitive division**: 17 objects are partitioned equally between 17 sets. How many objects are in each set?
(2) **Quotative division**: 17 objects are partitioned into sets, each containing 17 objects. How many sets are there?

In either case, the answer is obviously 1.

When the divisor is equal to the dividend, the quotient is always 1. In algebraic notation, for any (non-zero!) number a:

$$a \div a = 1.$$

8.4 Evaluating division

In this section, we describe various strategies for evaluating division with operands within the hundred range. The evaluation of division of multi-digit numbers is treated in Volume 2.

8.4.1 If you can count, you can divide

Consider the arithmetic expression:
$$12 \div 3.$$
Adopting a quotative division interpretation, its value is the number of sets obtained when 12 objects are partitioned into sets of 3 objects each.

Anyone who can count can evaluate this expression by drawing 12 objects,

partition them into sets of 3 items each,

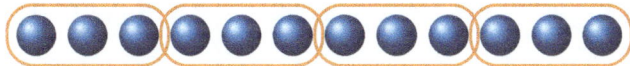

and count the number of sets—in this case, 4.

8.4.2 Evaluation by repeated addition

Consider the arithmetic problem:
$$56 \div 7 = \boxed{?}.$$
If we rephrase the problem in terms of quotative division, the question is: How many quotas of 7 are comprised in 56, or, how many times 7 make 56? One way of obtaining a solution is to skip count, starting from zero, adding 7 each time, until reaching the target number of 56:
$$7, 14, 21, 28, 35, 42, 49, \boxed{56}.$$
It remains then to count the number of skips. This evaluation strategy can be illustrated on a number line:

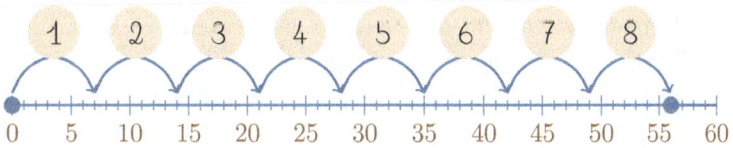

It takes 8 skips of 7 to reach 56, and therefore:
$$56 \div 7 = \boxed{8}.$$

8.4.3 Evaluation by repeated subtraction

Consider the same arithmetic problem:
$$56 \div 7 = \boxed{?}.$$

To partition 56 objects into quotas of 7, we can proceed as follows: We start with a set comprising 56 object. We take away 7 objects to form a first quota. We take away 7 more objects to form a second quota. We proceed this way, taking away 7 objects each time, until no objects remain. This procedure of repeated subtraction, or backward skip-count is illustrated below:

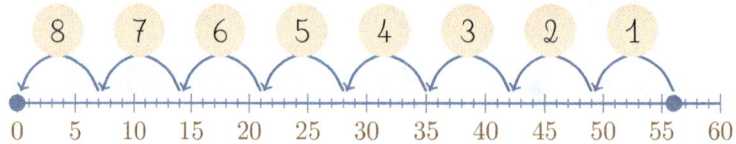

At the end, we count the number of backward skips from 56 to 0.

8.4.4 Chunking

Since the evaluation of division essentially requires finding an unknown factor, it is, to some extent, a task of reverse thinking. One approach to evaluating division is educated guesses.

Consider the following arithmetic problem:
$$84 \div 7 = \boxed{?}.$$

We can represent it with a word problem. For example:

> 84 bananas were divided evenly between 7 monkeys. How many bananas did each monkey get?

If you know your multiplication table, then you know that the answer has to be larger than 10, since 70 bananas suffice for each monkey to get 10.

Suppose we give 10 bananas to each monkey. Then, we have distributed 70 bananas out of 84. We have 14 bananas left, which we have to divide evenly between those 7 monkeys. 14 divided by 7 is 2, and therefore each monkey gets 2 more bananas. Summarizing, each monkey first receives 10 bananas, and then 2 more, hence,
$$84 \div 7 = \boxed{12}.$$

This evaluation strategy is known as **chunking** (i.e., dividing chunks until the total amount has been exhausted).

8.4.5 Division by 10

Just as there exists a particularly simple rule for multiplication by 10, there also exists a simple corresponding rule for division by 10.

Consider the equation:

$$70 \div 10 = \boxed{?}$$

which corresponds, as we know by now, to the unknown-factor equation:

$$\boxed{?} \times 10 = 70.$$

Since the multiplication of a number by 10 results in the digit 0 being appended to its right (see p. 94), it must be that division by 10 does the opposite: it results in the removal of the digit 0 from the ones column (at this point we limit ourselves to dividends that are multiples of 10, which ensures that the ones digits is indeed 0).

The rule whereby division by 10 entails the removal of the digit 0 from the ones column is useful to remember, but gives no hint for why it is valid. Consider again the above unknown-factor problem. Recalling that 70 is the decimal representation of 7 tens, we can reformulate the evaluation problem as follows:

> How many tens make 7 tens?

Rephrased in this way, the answer is obvious—7.

8.4.6 Division by 5

We already know how to multiply by 2 and how to divide by 10. These two skills can be combined to divide by 5.

Since $5 = 10 \div 2$, the laws of variation imply that for any number a:

$$a \div 5 = (a \div 10) \times 2.$$

Alternatively,

$$a \div 5 = (a \times 2) \div 10.$$

These two identities enable us to efficiently divide by 5.

When the dividend is a multiple of 10 it is easier to use the first identity—first divide by 10 and then multiply by 2. For example, to divide 130 by 5:

$$130 \div 5 = \underbrace{(130 \div 10)}_{13} \times 2 = 26.$$

When the dividend is not a multiple of 10 (but it is divisible by 5) we use the second identity—first multiply by 2 and then divide by 10. For example, to divide 75 by 5:

$$75 \div 5 = \underbrace{(75 \times 2)}_{150} \div 10 = 15.$$

Mathematical problems

Problem 8.1. Clovers are three-legged aliens.

(a) How many legs do 9 clovers have all together?
(b) Write down two division equations that are inverses of the product you evaluated in (a).
(c) Formulate a word problem for each division equation you wrote in (b). These word problems should relate to the word problem (a).

Problem 8.2. A package of candy was divided evenly among 5 children. Each child received 12 pieces of candy. One of the children decided to give up her share, and divided it evenly among her friends. How many pieces of candy did each child end up receiving?

Problem 8.3. It is given that $13 \times 154 = 2002$. Use this knowledge to evaluate
$$2002 \div 13 \quad \text{and} \quad 2002 \div 154.$$
Justify your answers.

Problem 8.4. It is given that $9 \div 0.03 = 300$. Use this knowledge to evaluate
$$18 \div 0.03.$$
Justify your answer. (Please note: you do not need to know what decimal fractions are in order to solve this problem!)

Problem 8.5. We have seen that division is not associative. For what values of a, b, and c does the identity:
$$(a \div b) \div c = a \div (b \div c)$$
holds? Try to formulate the most general answer.

Problem 8.6. Form a correct equation by placing the digits 6, 4, 9 and 3 in appropriate places:
$$\boxed{?}\,\boxed{?} \div \boxed{?} = \boxed{?}.$$
How many different solutions can you find?

Problem 8.7. Evaluate the following expressions:
$$385 \div 1 \quad \text{and} \quad 385 \div 385.$$
Justify your answers using both partitive division and quotative division arguments.

Problem 8.8. Evaluate the following expressions:
$$384 \times 3 \div 3 \qquad 384 \div 3 \times 3 \qquad 384 \times 2 \times 2 \div 4 \qquad 28 \div 4 \times 2.$$

Problem 8.9. Evaluate the following expressions:
$$(526 + 526) \div 2$$
$$(1423 + 1423 + 1423) \div 3.$$
Explain.

Problem 8.10. Fill in the appropriate signs, $<$, $>$, or $=$:
$$567 \div 3 \quad \underline{} \quad 524 \div 4$$
$$1001 \div 7 \quad \underline{} \quad 1001 \div 13.$$
Justify your answers.

Problem 8.11. Fill in the missing operations:

$$729 \div (9 \times 3) = (729 \div 9) \_\_ 3$$
$$729 \div (9 \div 3) = (729 \div 9) \_\_ 3.$$

Problem 8.12. Insert brackets in order to get a correct identity:

$$1024 \div 32 \div 16 = 1024 \times 16 \div 32.$$

Problem 8.13. Use the strategy presented in this chapter to evaluate the following expressions efficiently:

$$180 \div 5 \quad \text{and} \quad 215 \div 5.$$

Pedagogical problems

Problem 8.14. Teacher Cam presented the following word problem to her class:

> There are 3 cages in the zoo, and there are 4 gorillas in each cage. 120 bananas were distributed evenly to all the gorillas. How many bananas did each gorilla get?

She asked each child to represent the answer by an arithmetic expression.

Gili wrote: $120 \div (3 \times 4)$.
Alon wrote: $120 \div 3 \div 4$.

  (a) Are both arithmetic expressions equivalent?
  (b) Find a suitable illustration for each expression.
  (c) Is one of the answers more correct than the other?
  (d) Which mathematical rule presented in this chapter is reflected in this question?

Problem 8.15. Teacher Barbara presented the following word problem to her class:

> 8 children were divided into pairs. The children were given a total of 24 stickers, such that every pair of children got the same number of stickers. How many stickers were allotted to each pair of children?

She asked each child to represent the answer by an arithmetic expression.

Tal wrote: $24 \div (8 \div 2)$.
Anat wrote: $24 \div 8 \times 2$.

  (a) Are both arithmetic expressions equivalent?

(b) Find a suitable illustration for each expression.
(c) Is one of the answers more correct than the other?
(d) Which mathematical rule presented in this chapter is reflected in this question?

Chapter 9

Regrouping

In Chapters 4, 5, 7 and 8, we introduced the four operations of arithmetic. Our main focus was on interpreting each operation and understanding its properties. We also considered their evaluation, but only within the numerical ranges required in lower grades. When the operands are small, one can apply elementary evaluation methods, such as using manipulatives and skip counting. When the operands are large, such evaluation methods are impractical, and one has to resort to computational algorithms.

An **algorithm** is a recipe for solving a problem by following a step-by-step procedure that is guaranteed to terminate successfully after a finite number of operations. The word algorithm originates from an erroneous transliteration of the name of Persian mathematician *Al-Khawarizmi* (circa 780–850), who developed systematic methods for evaluating arithmetic operations. The suffix *arizmi* in his name was unconsciously replaced by the prefix *arithmi*, from the Greek *arithmetic*.

All computational algorithms have one thing in common: they are all based on the decimal numeral system. To understand the underlying principles of computational algorithms, a thorough understanding of the decimal system is required, and notably the notion of **regrouping** needs further elaboration.

9.1 Decimal units

Since the introduction of digits, we have at our disposal ten numerals representing numbers between zero and nine. In Chapter 2, we learned that our numeral system is based on the following convention: The number ten, which is the number following nine, is represented by the **bundling**, or **composition**, of ten ones into a new unit which we call a **ten**. One ten is equivalent to ten ones. This may sound like a tautology, but it isn't. The equivalence between ten ones and one ten is comparable to the equivalence between 60 seconds and 1 minute. Ones and tens are units among which there exists a fixed **conversion rate** of 10 to 1.

Just as we compose ten ones into one ten, we compose ten tens into a new unit which we call a **hundred**, and we compose ten hundreds into a new unit which we call a

thousand. The counting units of *one*, *ten*, *hundred* and *thousand* are called **decimal units**.

The conversion of ten decimal units by a new decimal unit is carried on recursively. Ten thousands make a decimal unit called simply a **ten-thousand**; ten ten-thousands make a **hundred-thousand**; ten hundred-thousands make a **million**; Ten millions are simply called a **ten-million**; ten of those make a **hundred-million**; ten hundred-millions make a **billion**; a thousand billions make a **trillion**; a thousand trillions make a **quadrillion**; there are names for even larger numbers.

 **Activity: Large numbers**

Children (and adults alike) find it difficult to grasp large numbers. Children recite "ten, a hundred, a thousand, a million, a billion" just as naturally as they recite "one, two, three".

To acquire a quantitative sense for large numbers it is necessary to encounter such numbers in concrete settings. The structured counting activity suggested in Chapter 2 (p. 20) can be performed with up to a thousand concrete objects (toothpicks, for example). A meticulous counting of a thousand objects can lead to a discussion about the space and time needed to gather ten-thousand toothpicks, a million toothpicks, and a billion toothpicks.

The MegaPenny project, by Kokogiac Media, is a website that offers a striking visualization of what would as many as one quintillion coins of one penny look like.

9.2 Place-value notation

In a decimal representation, every number is represented as a sum of decimal units. By convention, a given decimal unit may repeat at most nine times in the decimal representation of a number. There are exactly ten possibilities (between 0 and 9) for the number of repetitions of any decimal unit in the decimal representation of a number.

Each number is represented by a sequence of digits. By convention, the right-most digit represents the number of ones; the digit to its left represents the number of tens; the digit to its left represents the number of hundreds; and so on. The value of the decimal unit that each digit represents depends on the place of that digit within the numeral (see figure below), hence the name **place-value notation**.

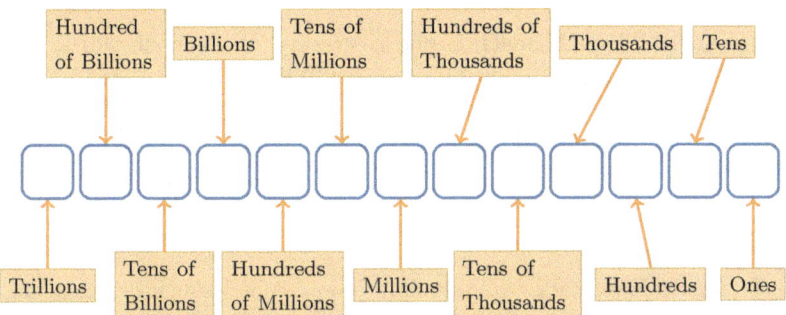

For example, the number

$$472314506975$$

represents, reading from right to left, the sum of: 5 ones, 7 tens, 9 hundreds, 6 thousands, 0 ten-thousands, 5 hundred-thousands, 4 millions, 1 ten-million, 3 hundred-millions, 2 billions, 7 ten-billions, and 4 hundred-billions.

Please note that with numbers as large as this one, it is hard to know at first glance what the value of the left-most digit, which is, in fact, the most significant digit (i.e., the digit that has the largest value) is. In other words, it is hard to estimate the number's magnitude without laboriously counting how many digits there are. In order to improve the readability of large numbers, it is customary (but not mandatory) to separate every three digits, from right to left, by a comma. Like this:

$$472, 314, 506, 975.$$

The commas partition the number into blocks: the ones block (the three right-most digits), the thousands block (the next three digits), the millions block, the billions block, and so on. Each one of these blocks is internally partitioned into ones, tens and hundreds. Separated into blocks, it is now easier to decipher this number as 472 billion, 314 million, 506 thousand and 975.

In the process of getting acquainted with the decimal numeral system, children are often asked to determine the value of a certain digit in the decimal representation of a number. A typical textbook question is:

What does the digit 5 represent in the number 4587?

The answer is that the digit 5 represents 5 hundreds: it quantifies the number of hundreds in the decimal representation of the given number.

Please note: The digit 5 represents the number of hundreds *in the decimal representation* of the number 4587. It does not represent the number of hundreds in the number itself, because there are, actually, 45.87 hundreds in 4587, and not 5 hundreds. One should not confuse properties of a number with properties of its decimal representation.

In conclusion, we note that our numeral system satisfies two important properties:

(1) There are no "illegal" sequences of digits: every finite sequence of digits represents a natural number.

(2) Every natural number has a unique representation.

9.2.1 Zero as a place holder

When the decimal representation of a number does not include certain decimal units but contains units of higher value, the place-value notation necessitates the use of a place holder. As explained in Chapter 2, the digit 0 was naturally chosen for this task. For example, the number made up of 5 hundreds and 7 ones is written as 507. The fact that the decimal representation contains no tens must be indicated explicitly.

Consider the following number:

$$000237.$$

It represents 7 ones, 3 tens, 2 hundreds, 0 thousands, 0 ten-thousands and 0 hundred-thousands. The three left-most zeroes are in fact superfluous. They can be omitted, because they do not hold the place for any digit of higher-value. A sequence of zeros in the decimal representation of a number after which no other digits follow is called **trailing zeroes**.

Should we then confuse children by exposing them to numeral with trailing zeroes? There are at least three good reasons to consider such numerals:

(1) Children encounter trailing zeroes in digital representations of time, as in 03:24, or in the representation of dates, as in 03/08/09.

(2) Trailing zeroes occur in decimal fractions, where it is the right-most zeroes that are superfluous, as in the number 0.56000.

(3) Trailing zeroes can be useful, as in vertical addition. By appending trailing zeroes, we can always create a situation in which the two addends have the same number of digits. It allows us to easily align them one under the other.

It should be emphasized that the inclusion of trailing zeroes is by no means a mistake.

9.3 Regrouping

Every two decimal units are related by a fixed conversion rate. The conversion can go either way:

(1) **Composing**: The replacement of 10 identical decimal units by one decimal unit whose value is 10 times larger.
(2) **Decomposing**: The replacement of 1 decimal unit by 10 decimal units, the value of each being 10 times smaller.

The process of changing the representation of a number by either composing or decomposing decimal units is called **regrouping**. Regrouping is at the heart of all computational algorithms. In addition and multiplication, we often have to compose decimal units; whereas in subtraction and division, we often have to decompose them.

9.3.1 Non-standard decimal representations

Imagine a pile of 325 objects (we use the word *pile* as a metaphor for objects that are not organized in any structured way). The first stage in forming the decimal representation of this quantity is to compose tens of objects, until there are fewer than ten scattered objects left. At the end of this stage, we have 32 tens of objects and 5 scattered objects.

If we allowed ourselves to use a decimal notation in which the same decimal unit can be repeated an unlimited number of times, we could represent the number of objects as follows:

Our numeral system, however, limits each decimal unit to repeat at most 9 times. Thus we need to compose hundreds by grouping together tens of tens, until fewer than 10 tens remain. At the end of this stage, we have 3 hundreds of objects, 2 tens of objects, and 5 scattered objects, which can be represented in standard decimal representation:

If the need arises, however, we can modify the numeral by regrouping decimal units without changing the value that that numeral represents. For example, we can decompose one hundred into 10 tens, and obtain the following representation:

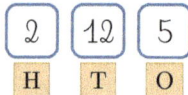

If we further decompose one ten into 10 ones, we obtain:

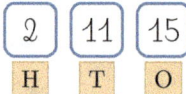

Let's emphasize: the combination of decimal units:

> 2 hundreds, 11 tens and 15 ones

represents the *same number* as the combination of decimal units:

> 2 hundreds, 12 tens and 5 ones,

and they both represent the *same number* as:

> 3 hundreds, 2 tens and 5 ones.

It is only this last combination of decimal units that complies with our conventional numeral system. In the next few chapters, we will make extensive use of combinations of decimal units that do not limit the number of repetitions of each decimal unit. We will refer to these as **non-standard decimal representations**.

9.4 Numeral systems and complexity

In this last section, we will examine our numeral system from the vantage point of a mathematical field called **complexity theory**.

Before introducing the decimal system in Chapter 2, we introduced a more primitive numeral system, a **unary** system, in which every number is represented by a set of symbols that match exactly the set of objects they represent. In a unary system, the number sequence looks like this:

Since there is a one-to-one correspondence between the quantity and its numeral representation, we need 10 times as many symbols to represent a number that is 10 times larger. Think of an extreme example—representing the total number of particles that make up the universe. Scientists estimate that this number is a 70-digit number. How can such a number be represented using a unary numeral system? There is not enough matter in the universe to create such a representation!

With that in mind, we see that the first advantage of the decimal numeral system is **succinctness**: the representation is much more compact than the quantity it represents. For example, a number that is 10 times larger than another requires only 1 additional digit to represent it. Likewise, a number that is a million times larger than another requires only 6 additional digit to represent it. The mathematical terminology that describes this situation is that the size of the decimal representation increases **logarithmically** with the size of the quantity that it represents. Put simply, when we change a quantity *multiplicatively* (e.g., we replace a number by another that is a million times larger), the number of digits in the numeral representation changes only *additively* (6 more digits). Isn't our numeral system a most wonderful invention?

But as the saying goes, there is no such thing as a free lunch. The price we have to pay for having adopted a succinct and powerful numeral system is that we represent quantities using a *code* that constantly requires deciphering. Acquiring familiarity with this code is perhaps the major goal of mathematics in the lower grades.

What do we mean by a *code*? Borrowing terminology from the field of cryptology, when children learn to count and represent their result as a numeral, as in:

$$||||||||||||||||||| \;\; \rightarrow \;\; 18,$$

they are performing some kind of **encoding**. Conversely, when children convert a numeral into a quantity of objects, they perform some kind of **decoding**.

In Chapters 4, 5, 7 and 8, we asserted that whoever can count can also add, subtract, multiply and divide. In all four cases, evaluation is based on the following sequence of steps:

(a) **Decoding**: Convert the two operands, which are given in the form of decimal numerals, into a set of symbols. That is, the data, which is given in decimal representation is converted into a unary representation.

(b) **Calculation**: Given the unary representations of the two operands, the calculations amount to simple operations (e.g., group together two sets of symbols) and counting.

(c) **Encoding**: Re-express the outcome of the counting in decimal representation.

In other words, all the strategies of "if you can count, you can" are based on a sequence of decoding-calculation-encoding procedures. The major drawback of these strategies is that the operations (e.g., counting symbols) are performed in unary representation. Hence, the number of basic operations we have to perform is as large as the quantities being manipulated. Once again, adopting the terminology of complexity theory, the complexity of the calculation scales **linearly** with the magnitude of the data. The same operation with numbers that are a million times larger will take a million times longer!

Evaluation algorithms (two of which we will learn about in the next two chapters) do not require decoding the data into a unary representation. The algorithms operate directly on the decimal representation. The benefit is immense: the number of basic operations we have to perform scales logarithmically (i.e., very slowly) with the magnitude of the data. The same operation with numbers a million times larger only takes a little longer to evaluate. Every child knows the cost: in order to be able to evaluate arithmetic expressions with large operands, we have to acquire a proficiency in the performance of sequences of mechanical operations. Often children do not understand how those methods work and why they yield the correct answer.

Mathematical problems

Problem 9.1. By how many digits does the decimal representation of a number grow when we multiply it by a billion?

Problem 9.2. Express the number 1305 using five different non-standard representations. For example:

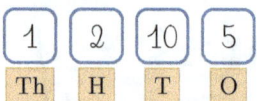

Chapter 10

Addition of Multi-Digit Numbers

In Chapter 4, we introduced addition. We focused on its various interpretations, its properties, and on proficiency in evaluating sums within the range of 20. In this chapter, we will combine this knowledge with the notion of regrouping introduced in Chapter 9. We will learn how to integrate addition within the range of 20 and regrouping to evaluate sums with arbitrarily large addends.

10.1 Addition without regrouping

Consider the following expression:

$$435 + 252.$$

Its evaluation might seem difficult to a child who has only learned to evaluate sums within the range of 20. But if we recall the interpretation of addition as joining together two quantities, and we know how to decompose a multi-digit number into a sum of decimal units, then we can rephrase the problem as follows:

> What is the outcome of joining together 4 hundreds, 3 tens and 5 ones, and 2 hundreds, 5 tens and 2 ones?

Forget momentarily that *hundreds*, *tens* and *ones* are decimal units among which there exists a fixed conversion rate. If we treated them as three independent nouns, such as *apples*, *rabbits*, and *hats*, we could have asked:

> What is the result of joining together 4 apples, 3 rabbits and 5 hats, and 2 apples, 5 rabbits, and 2 hats?

This question can be answered by first graders. We face three separate summations, all within the range of 10. The result is 6 apples, 8 rabbits and 7 hats.

Reverting apples to hundreds, rabbits to tens and hats to ones:

The result of joining together 4 hundreds, 3 tens and 5 ones, and 2 hundreds, 5 tens and 2 ones is 6 hundreds, 8 tens and 7 ones.

Read the answer. It is a sum of decimal units, which we know how to represent as a numeral: 687. Thus,

$$435 + 252 = \boxed{687}.$$

We can make the situation concrete in another way: Suppose that the objects quantified by the two addends were dollars (or any other currency, or even better, toy money). Money is a useful manipulative because the currency tallies with the decimal system: there are banknotes of \$1, \$10, and \$100 (some countries have banknotes that correspond to even larger decimal units). In fact, banknotes are halfway between the concrete and the abstract. They are concrete in the sense of being objects that can be manipulated and exchanged. They are abstract in the sense that a \$100 bill is not really a bundle of a hundred \$1 bills. To correctly simulate our decimal notation, be sure not to use \$5, \$20 and \$50 bills.

We add 435 and 252 by joining together two piles of banknotes as follows:

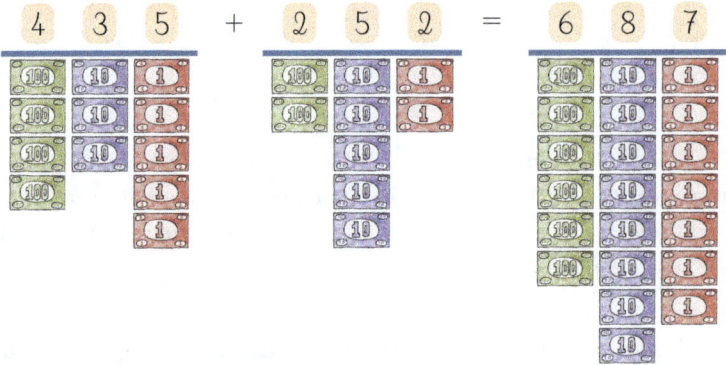

This figure clarifies that in this case the addition of 3-digit numbers only entails the separate addition of same decimal unit, each within the range of 10.

Please note that we applied the associative property of addition. If addition weren't associative, we would not have been able to join hundreds with hundreds, tens with tens and ones with ones.

10.2 Vertical addition without regrouping

The lesson from the previous section is that the addition of multi-digit numbers can be broken down into a sequence of additions of single-digit numbers. Our next goal is to automate this process in order to add rapidly and efficiently, without having to imagine apples, rabbits, and hats, and without having to manipulate banknotes.

Consider the following expression:

$$23,506 + 5,220.$$

The first addend comprises 2 ten-thousands, 3 thousands, 5 hundreds, 0 tens and 6 ones. The second addend comprises 5 thousands, 2 hundreds, 2 tens and 0 ones. As in the previous example, we join together same decimal units, and obtain 2 ten-thousands, 8 thousands, 7 hundreds, 2 tens and 6 ones. That is:

$$23,506 + 5,220 = \boxed{28,726}.$$

And suppose we wanted to add even larger numbers:

$$10,325,004,302 + 546,421,903,477.$$

In principle, this sum is not harder to evaluate than the previous ones. However, because there are so many digits, our eyes begin to dart between the two numbers, and it is difficult to know which digit in the first addend matches which digit in the second addend.

To cope with this difficulty an effective notational system was invented—**vertical addition**. By writing the two addends one on top of the other, as below, digits that represent the same decimal unit are aligned vertically.

```
        1 0 3 2 5 0 0 4 3 0 2
    +   5 4 6 4 2 1 9 0 3 4 7 7
        ─────────────────────
```

It is essential that both addends be *right-aligned*. The use of grid paper facilitates the task of keeping a precise vertical alignment, especially when adding very large numbers.

All that needs to be done is to add ones and ones, tens and tens, etc. The result of each summation is written under the two addends, aligned with the corresponding digits of the addends:

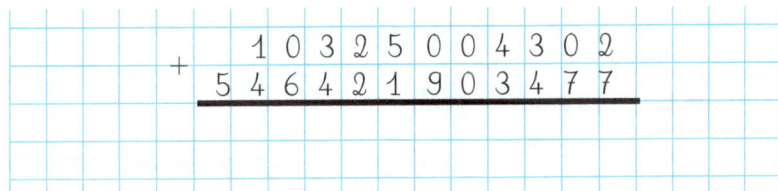

In principle, we may sum up the various decimal units in any order, relying on both commutative and associative properties of addition. We can start adding up the ones first and proceed from right to left. We can also start with the left-most digit, or if we please, proceed in any other order. The order of summation is significant when the sum comprises decimal units that repeat 10 times or more. In this case, regrouping is needed, and the algorithm has to be amended accordingly (see the next section). As we will see, addition with regrouping is more convenient if we start summing up the decimal units from right to left, i.e., from smallest to largest.

10.3 Addition with regrouping

Consider the following expression:

$$436 + 287.$$

Following the steps described in the previous section, we join together 4 hundreds, 3 tens and 6 ones with 2 hundreds, 8 tens and 7 ones. We obtain 6 hundreds, 11 tens and 13 ones. This time the result is not in standard decimal notation. If we use a non-standard decimal notation as we did in Chapter 9, we can write the sum as follows:

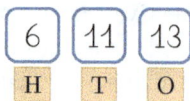

This answer is correct; only written in non-standard notation. Next, we have to standardize the representation of the sum. This is done by regrouping the decimal units such that each unit repeats at most 9 times.

Let's start with the ones. There are 13 of them. We take 10 of them to compose a ten. We are left with just 3 ones. The newly composed ten adds to the 11 tens that were there at the start, yielding 12 tens. In non-standard notation:

$$\boxed{6}\;\boxed{11}\;\boxed{13} \;=\; \boxed{6}\;\boxed{12}\;\boxed{3}$$
$$\;\text{H}\;\;\;\text{T}\;\;\;\text{O}\quad\quad\;\text{H}\;\;\;\text{T}\;\;\;\text{O}$$

Consider next the tens. There are 12 of them. We take 10 of them to compose a hundred. We are left with just 2 tens. The newly composed hundred adds to the 6 hundreds that were there at the start, yielding 7 hundreds. In our non-standard notation:

$$\boxed{6}\;\boxed{12}\;\boxed{3} \;=\; \boxed{7}\;\boxed{2}\;\boxed{3}$$
$$\;\text{H}\;\;\;\text{T}\;\;\;\text{O}\quad\quad\;\text{H}\;\;\;\text{T}\;\;\;\text{O}$$

After two regrouping steps, we obtain a number in standard decimal notation: 723. Please note: in the transition from 6|11|13 to 6|12|3, and then to 7|2|3, we do not change the number. We only modify its representation.

Once again, we can use currency to illustrate this evaluation process. First, we add ones, tens and hundreds separately, in the following way:

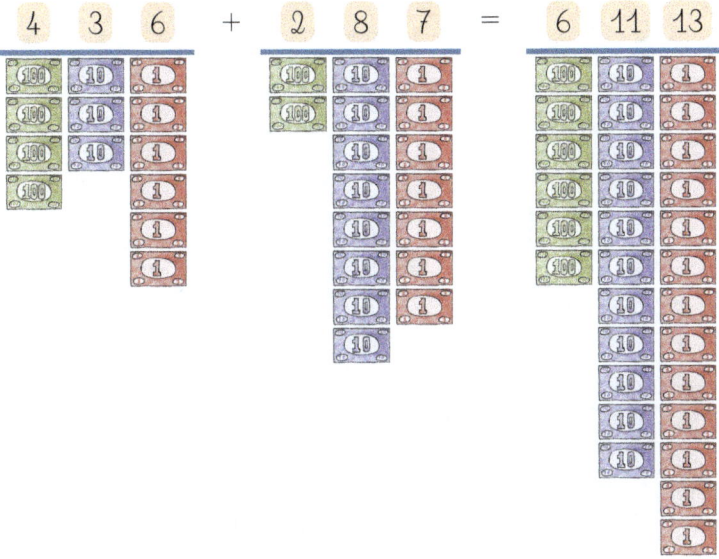

Since there are more than 10 ones, we exchange 10 ones for 1 ten:

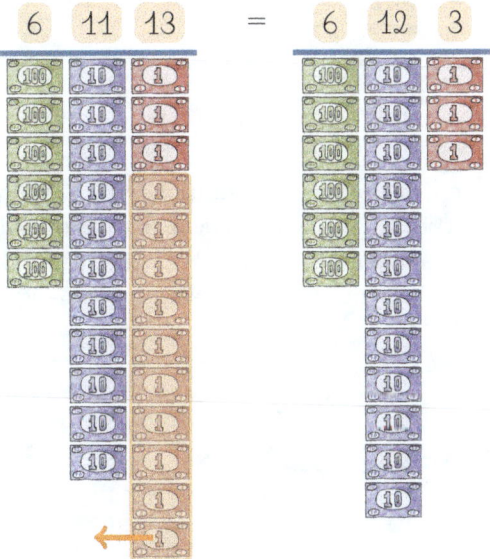

Fewer than 10 ones remain, but there are more than 10 tens so we exchange 10 tens for 1 hundred:

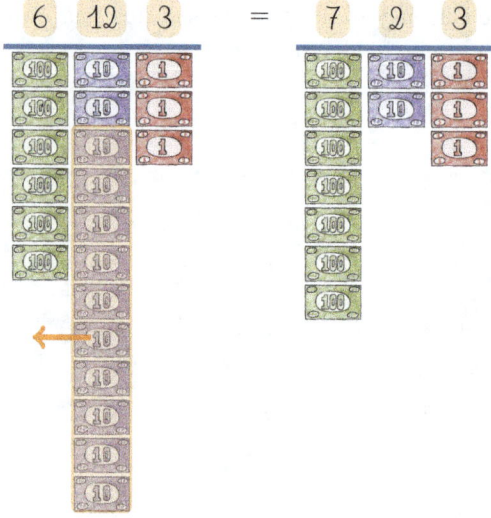

Now every type of banknote repeats at most 9 times. Thus, we obtained the standard representation of the solution: 7 hundreds, 2 tens, and 3 ones, or, 723.

Please note that when we use currency as manipulatives, regrouping takes the form of an exchange of banknotes. Most children know from daily experience that when you exchange 10 ones for a ten or 10 tens for a hundred, the amount of money does not change. We should capitalize on this understanding, as it is not always clear to children that regrouping is only a change in representation, and not a change in value.

10.4 Vertical addition with regrouping

Just as in addition without regrouping, in addition with regrouping it is convenient to align the two addends vertically. In the first stage, we add each decimal unit separately, adding ones and ones, tens and tens, and so on. We write the result of each summation in the appropriate column of the sum, even if this sum is a two-digit number. For example:

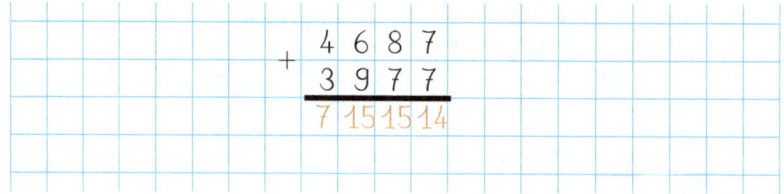

This is the result in non-standard notation. Next, we regroup the decimal units until each decimal unit repeats at most 9 times. Starting from the right-most digit, we perform the following steps:

$$\boxed{7}\ \boxed{15}\ \boxed{15}\ \boxed{14} \quad = \quad \boxed{7}\ \boxed{15}\ \boxed{16}\ \boxed{4}$$
$$\text{Th}\quad \text{H}\quad \text{T}\quad \text{O} \qquad\qquad \text{Th}\quad \text{H}\quad \text{T}\quad \text{O}$$

and then

$$\boxed{7}\ \boxed{16}\ \boxed{6}\ \boxed{4} \quad = \quad \boxed{8}\ \boxed{6}\ \boxed{6}\ \boxed{4}$$
$$\text{Th}\quad \text{H}\quad \text{T}\quad \text{O} \qquad\qquad \text{Th}\quad \text{H}\quad \text{T}\quad \text{O}$$

This method works, but it is awkward. The use of a non-standard notation—*a very useful didactic tool*—involves notational burdens, like framing numbers in boxes in order to keep track of which digits belong to which unit.

The standard algorithm obviates the use of non-standard notation, circumventing such burdens. We demonstrate this using the same example. We start by writing the two addends vertically, aligning matching digits. We then proceed, step-by-step, in the following order:

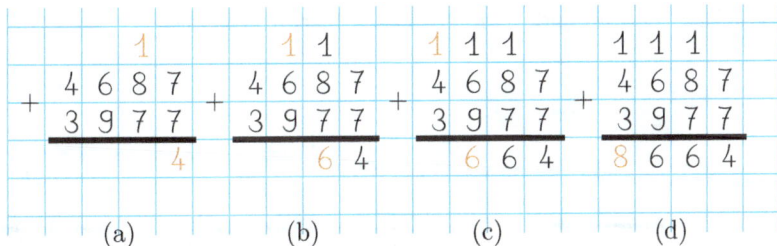

(a)   (b)   (c)   (d)

(a) **Add ones**: 7 ones from the first addend plus 7 ones from the second addend make 14 ones. Since there are more than 9 ones, we compose a ten, thus obtaining 1 ten and 4 ones. We record the 4 ones by writing the digit 4 in the ones column of the sum. We also have 1 ten, which we have "put aside" until we sum the tens. It is customary to represent this ten by writing the digit 1 in the column allocated to tens, above the first addend. This 1 is called a **carry**, meaning that it represents a ten that was carried over from summing up ones.

(b) **Add tens**: 8 tens from the first addend plus 7 tens from the second addend plus 1 ten carried from the ones, make together 16 tens. Since there are more than 9 tens, we compose a hundred, thus obtaining 1 hundred and 6 tens. We record the 6 tens by writing the digit 6 in the tens column of the sum. We also have 1 hundred, which we carry to the hundreds column, above the first addend.

(c) **Add hundreds**: 6 hundreds from the first addend plus 9 hundreds from the second addend plus 1 hundred carried from the tens, make together 16 hundreds. Since there are more than 9 hundreds, we compose a thousand, thus obtaining 1 thousand and 6 hundreds. We record the 6 hundreds by writing the digit 6 in the hundreds column of the sum. We also have 1 thousand, which we carry to the thousands column, above the first addend.

(d) **Add thousands**: 4 thousands from the first addend plus 3 thousands from the second addend plus 1 thousand carried from the hundreds, make together 8 thousands. We record the 8 thousand by writing the digit 8 in the thousands column of the sum.

 Comments

(1) Please note how important it is to be proficient in evaluating addition within the range of 20.

(2) It may be that the number of digits in the sum is larger than the number of digits of either addend, as in the case illustrated below:

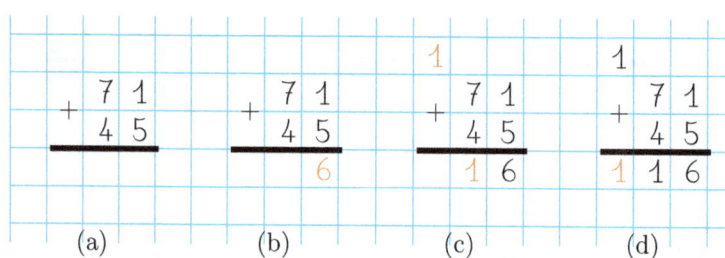

Up to the summation of tens, the stages are the same as in the previous example. The difference is only in the last stage, where we have to compose a hundred from 10 tens. Since neither addends contain hundreds, we have to allocate a hundreds column. We represent the hundred carried from the tens in the hundreds column above the two addends. Finally, we sum up the hundreds, which only comprise the carry.

Mathematical problems

Problem 10.1. Answer each of the following questions with minimum effort. Justify your answers in detail.

(a) What is the ones digit in $4,657 + 32,115$?
(b) What is the tens digit in $45,564,712 + 3,456,549$?
(c) What is the hundreds digit in $34,562 + 546,219$?
(d) What is the hundreds digit in $43,587 + 1,236$?
(e) What is the millions digit in $345,789 + 995,549$?

Problem 10.2. Find an efficient way to evaluate the following sum:

$$999,999 + 2,134.$$

Problem 10.3. In the following problem, some of the digits were accidentally erased:

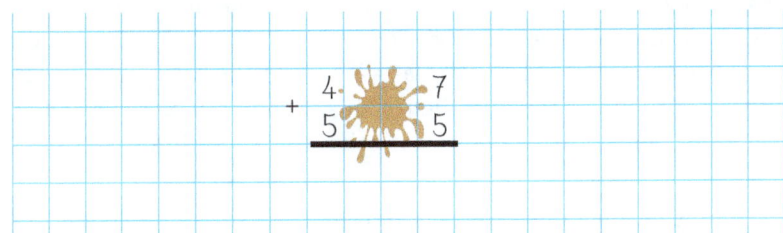

(a) Is it possible to know what is the ones digit of the sum?
(b) Can the tens digit of the sum be any digit?
(c) Which digits may appear in the thousands digit of the sum?
(d) What is the largest possible value of the sum? What is its smallest possible value?

Pedagogical problems

Problem 10.4. Analyze how a child could have reached the following erroneous result:

$$1,235 + 745 = 8,685.$$

How would you help this child not to repeat this error in the future?

Problem 10.5. Teacher Jan asked her class to evaluate the following sum:

$$912 + 164 + 586 + 638.$$

Tal presented the following solution:

$$910 + 160 + 590 + 640$$
$$900 + 200 + 600 + 600$$
$$1000 + 100 + 1000 + 200 = \boxed{2300}.$$

(a) Explain Tal's strategy.
(b) Is this strategy always appropriate?
(c) How would you compare this strategy with vertical addition? Which is easier? Which is more systematic?

Problem 10.6. Teacher Bob asked his class to evaluate mentally the following sum:

$$368 + 156.$$

Gili: I added up the hundreds first, and got 400. Then I added up the tens, and got 110, so that in total I already had 510. Finally, I added up the ones and got 14, so the result is 524.

Alon: First, I calculated only the ones in the sum of each column and ignored regroupings. This way I had 4 in the ones digit, 1 in the tens digit, 4 in the hundreds digits, and 414 in total. Second, I saw that the sum of the ones digit is larger than 10 and that the sum of the tens digit is larger than 10, so I added 1 to the tens digit, and I added 1 to the hundreds column, and got 524.

(a) What do you think of these two strategies?
(b) Do these strategies always work? Are they efficient?
(c) Would you recommend these strategies to children?

Chapter 11

Subtraction of Multi-Digit Numbers

In Chapter 5, we introduced subtraction. We focused on its various interpretations, its properties, and on proficiency in evaluating subtraction within the range of 20. In this chapter, we combine this knowledge with the notion of regrouping introduced in Chapter 9. We will learn how to integrate subtraction within the range of 20 and regrouping to evaluate subtraction with arbitrarily large operands. Please note the strong analogy between this introduction and the introduction to Chapter 10, which addressed the evaluation of addition with arbitrarily large addends.

11.1 Subtraction without regrouping

Consider the arithmetic expression:
$$648 - 216.$$
Its evaluation might seem difficult to a child who has only learned to evaluate differences within the range of 20. But if we recall the interpretation of addition as taking away, and we know how to decompose a multi-digit number into a sum of decimal units, then we can rephrase the problem as follows:

> What remains if we take away 2 hundreds, 1 ten and 6 ones from 6 hundreds, 4 tens and 8 ones?

We repeat the steps taken in Chapter 10 in the context of addition (p. 129). Forget momentarily that *hundreds*, *tens* and *ones* are decimal units among which there exists a fixed conversion rate. If we treated them as three independent nouns, like say, *apples*, *rabbits*, and *hats*. We could ask instead:

> What remains if we take away 2 apples, 1 rabbit and 6 hats from 6 apples, 4 rabbits and 8 hats?

This question can be answered by first graders. They would have to solve three separate problems: one for apples, one for rabbits and another for hats. All three problems require the evaluation of a difference within the range of 10. The answer is: 4 apples, 3 rabbits and 2 hats.

Replacing apples with hundreds, rabbits with tens and hats with ones, we arrive at the following solution:

> If we remove 2 hundreds, 1 ten and 6 ones from 6 hundreds, 4 tens and 8 ones, there remain 4 hundreds, 3 tens and 2 ones.

Read the answer. It is a sum of decimal units, which we know how to represent as a numeral: 432. Thus,

$$648 - 216 = \boxed{432}.$$

In this particular example, no regrouping was needed. We were able to break down the evaluation into three separate tasks, one for each decimal unit.

We can use currency to illustrate the evaluation of this difference.

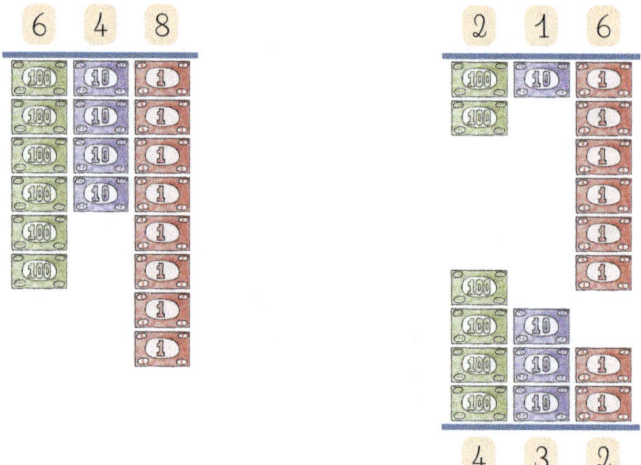

As in Chapter 10, we only use banknotes of $1, $10 and $100. On the left, we display the minuend. On the right, we separate the minuend into two amounts, one of which represents the subtrahend. The difference is the complement: 4 hundreds, 3 tens and 2 ones.

11.2 Vertical subtraction without regrouping

The lesson from the previous section is that subtraction of multi-digit numbers can be broken down into separate subtractions of single-digit numbers. Our next goal is to facilitate this process, without imagining apples, rabbits, and hats, or manipulating banknotes.

Just as in addition, it is convenient to write the two operands one above the other, such that digits representing the same decimal unit are aligned vertically. We allocate a row under both operands for the difference. The key difference between subtraction and addition is that subtraction is not commutative. Therefore, the order of the operands is significant. The convention is to write the minuend above the subtrahend, as follows:

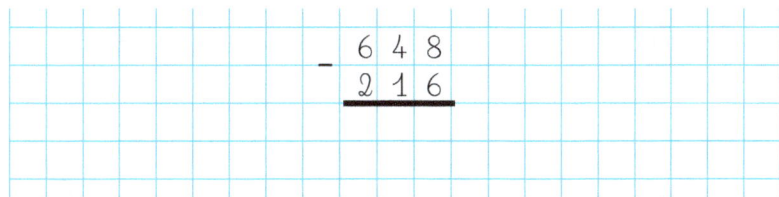

To evaluate the difference, we subtract the ones digit of the subtrahend, 6, from the ones digit of the minuend, 8, and write the difference, 2, in the ones column of the difference. We repeat the same procedure for the tens and the hundreds, as shown below:

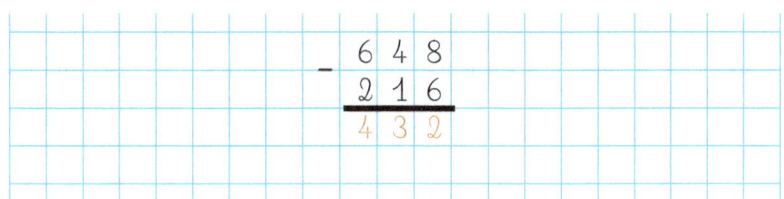

Please note that since this evaluation does not necessitate regrouping, it does not matter in which order we proceed.

 Validation

Any subtraction can be validated using addition (see Chapter 5, p. 66). If it were true that 432 is the difference between 648 and 216, it is would necessary imply that

$$216 + 432 = 648.$$

This identity can be checked by evaluating the left-hand side. We can use either vertical addition, or any other evaluation method.

11.3 Vertical subtraction with regrouping

Consider the arithmetic expression:

$$646 - 218.$$

What is its value? As in the previous section, we rephrase the problem as follows:

> What remains if we take away 2 hundreds, 1 ten and 8 ones from 6 hundreds, 4 tens and 4 ones?

Once again, let's replace *hundreds*, *tens* and *ones* by *apples*, *rabbits*, and *hats*, and ask instead:

> What remains if we take away 2 apples, 1 rabbit and 8 hats from 6 apples, 4 rabbits and 6 hats?

We run into a problem: how can we remove 8 hats if there are only 6 hats? It is here that we must exploit the fact that hundreds, tens and ones are not arbitrary nouns, but decimal units that can be composed and decomposed. A ten can be decomposed into 10 ones, and a hundred can be decomposed into 10 tens.

11.3.1 First regroup and then evaluate

When we evaluate addition, we can always join together the same decimal units. Depending on the addends, it may happen that the resulting sum is not in standard notation, and must be regrouped. When we evaluate subtraction, the opposite occurs. We may need to regroup the operands *prior* to subtracting the same decimal units.

Consider once again the difference between 646 and 218. The decimal representation of the minuend comprises 6 ones, whereas the decimal representation of the subtrahend comprises 8 ones. It looks like there are not enough ones to be removed. We can, however, regroup the minuend and decompose a ten into 10 ones. We obtain a different representation of the minuend:

$$\boxed{6}\ \boxed{4}\ \boxed{6} \ = \ \boxed{6}\ \boxed{3}\ \boxed{16}$$
$$\ \text{H}\ \ \text{T}\ \ \text{O} \qquad\ \text{H}\ \ \text{T}\ \ \text{O}$$

We may now subtract, ones from ones, tens from tens and hundreds from hundreds:

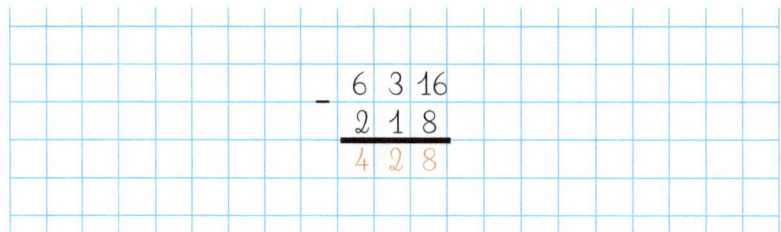

This solution strategy involves two distinct steps:

(1) **Regroup**: Represent the minuend such that each decimal unit repeats at least as many times as in the subtrahend.

(2) **Evaluate**: Subtract one decimal unit at a time.

This solution strategy has many didactic advantages. Regrouping and subtraction are two inherently different operations, a fact that is often obscured by the technicalities of the standard algorithm.

11.3.2 The standard algorithm

The only disadvantage of the *first regroup and then evaluate* approach is notational (compare with a similar statement in Chapter 10). The standard subtraction algorithm interlaces regrouping and evaluation. It is a more compact notation, but not quite as clear.

The standard subtraction algorithm applied to the evaluation of the difference between 646 and 218 is as follows:

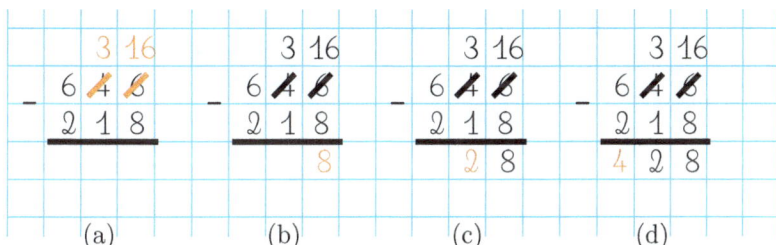

(a) **Regroup**: Decompose 1 ten into 10 ones. As a result 3 tens remain and there are 16 ones. Denote this change by crossing out the original ones and tens digits, and write the new digits in a row above the minuend.

(b) **Evaluate ones**: Subtract 8 ones from 16 ones. Remainder: 8 ones. Write the digit 8 in the ones column of the difference.

(c) **Evaluate tens**: Subtract 1 ten from 3 tens. Remainder: 2 tens. Write the digit 2 in the tens column of the difference.

(d) **Evaluate hundreds**: Subtract 2 hundreds from 6 hundreds. Remainder: 4 hundreds. Write the digit 4 in the hundreds column of the difference.

11.3.3 Multiple regroupings

Consider the arithmetic expression:

$$643 - 276.$$

We are going to evaluate it using the two approaches described above: the *first regroup and then evaluate* approach and the standard algorithm. I encourage you to decide for yourself which way works best.

11.3.3.1 First regroup and then evaluate

Examine both minuend and subtrahend: The ones digit of the subtrahend is larger than the ones digit of the minuend, so we need to regroup the minuend and decompose a ten into 10 ones. The tens digit of the subtrahend is also larger than the tens digit of the minuend, so we also need to decompose a hundred into 10 tens. Using the non-standard decimal notation, we regroup the minuend in two steps:

6	4	3	=	6	3	13
H	T	O		H	T	O

and

6	3	13	=	5	13	13
H	T	O		H	T	O

With this new representation of the minuend, we may evaluate the difference, digit after digit, as follows:

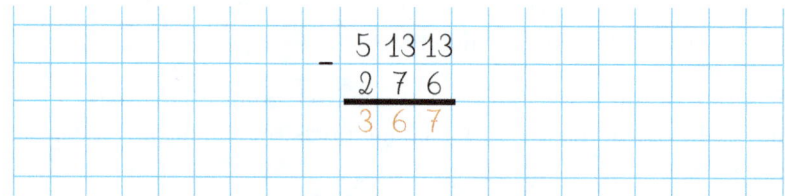

11.3.3.2 The standard algorithm

The standard algorithm interlaces regrouping and evaluation as follows:

(a) **Regroup**: Decompose 1 ten into 10 ones. 3 tens remain and there are 13 ones. Denote this change by crossing out the original ones and tens digits, and writing the new digits above the minuend.

(b) **Evaluate ones**: Subtract 6 ones from 13 ones. Remainder: 7 ones. Write the digit 7 in the ones column of the difference.

(c) **Regroup**: Decompose 1 hundred into 10 tens. 5 hundreds remain and there are 13 tens. Denote this change by crossing out the original tens and hundreds digits, and writing the new digits above the minuend.

(d) **Evaluate tens**: Subtract 7 tens from 13 tens. Remainder: 6 tens. Write the digit 6 in the tens column of the difference.

(e) **Evaluate hundreds**: Subtract 2 hundreds from 5 hundreds. Remainder: 3 hundreds. Write the digit 3 in the hundreds column of the difference.

Consider another expression whose evaluation requires multiple regrouping:

$$5,003 - 427.$$

If we were to implement the *first regroup and then evaluate* approach, we would observe that the decimal representation of the minuend contains fewer ones, tens and hundreds than the decimal representation of the subtrahend. To represent the minuend in such a way that it contains sufficiently many ones, we need to decompose a ten. But there are no tens to decompose. What can we do? Decompose a hundred into 10 tens? There are no hundreds either so the only option is to decompose a thousand into 10 hundreds. We proceed from left to right, as follows:

(a) Decompose a thousand into 10 hundreds.
(b) Decompose a hundred into 10 tens.
(c) Decompose a ten into 10 ones.

Those three steps are shown below using non-standard notation:

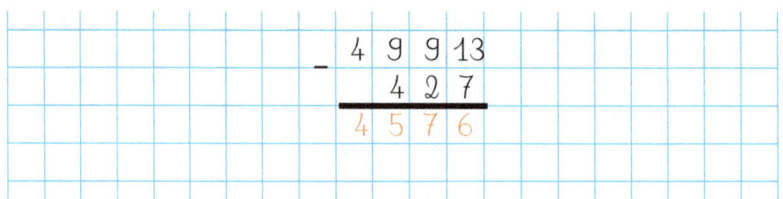

We may now proceed with the evaluation, one digit at a time:

$$\begin{array}{r} 4\;9\;9\;13 \\ -\;4\;2\;7 \\ \hline 4\;5\;7\;6 \end{array}$$

In contrast, the vertical subtraction algorithm is displayed below:

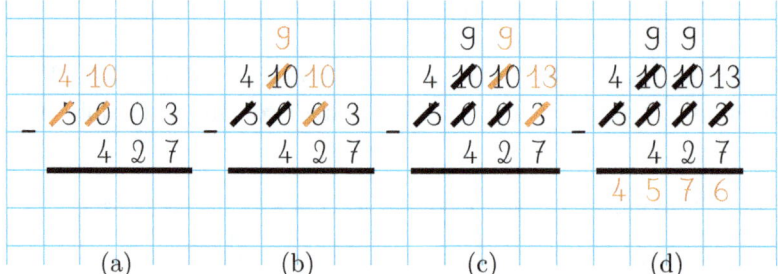

(a)   (b)   (c)   (d)

11.4 The French algorithm

The standard subtraction algorithm is not the only one that is both systematic and efficient. In France, for example, they use an algorithm that is different from the one used in most of the world. The French algorithm is instructive even if you do not intend to use it, because it takes advantage of the fact that the difference does not change if both minuend and subtrahend vary by the same amount in a clever way.

We will exemplify the French algorithm by evaluating an expression already evaluated:

$$643 - 276.$$

The French algorithm consists of the following steps:

Subtraction of Multi-Digit Numbers

```
        10            10 10         10 10
  6 4 3       6 4 3       6 4 3       6 4 3
- 2 7 6     - 2 7 6     - 2 7 6     - 2 7 6
                1           1 1         1 1
                  7           6 7     3 6 7

   (a)         (b)         (c)         (d)
```

(a) Write both operators in vertical alignment as in the standard algorithm.

(b) **Evaluate ones**: We need to subtract 6 ones from 3 ones. In the standard algorithm, we decompose a ten into 10 ones. Instead, in the French algorithm, we add 10 to both minuend and subtrahend: we add 10 ones to the minuend and 1 ten to the subtrahend. The 10 ones are written in the ones column above the minuend, whereas the 1 ten is written in the tens column under the subtrahend.

Subtract 6 ones from 10 + 3 ones. Remainder: 7 ones. Write the digit 7 in the ones column of the difference.

(c) **Evaluate tens**: We need to subtract 7 + 1 tens from 4 tens. Again, we vary both minuend and subtrahend by the same amount. This time we add 10 tens to the minuend and 1 hundred to the subtrahend.

Subtract 7 + 1 tens from 10 + 4 tens. Remainder: 6 tens. Write the digit 6 in the tens column of the difference.

(d) **Evaluate hundreds**: Subtract 2 + 1 hundreds from 6 hundreds. Remainder: 3 hundreds. Write the digit 3 in the hundreds column of the difference.

Mathematical problems

Problem 11.1. Evaluate the following expression using vertical subtraction:

$$70,000,013 - 14.$$

Propose a more effective solution strategy.

Problem 11.2. In the following problem, some of the digits were accidentally erased:

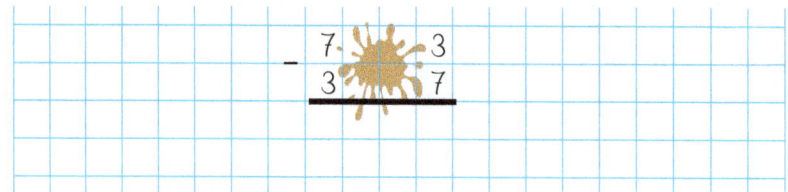

(a) Is it possible to know what is the ones digit of the difference?
(b) Can the difference be less than $4,000$?
(c) Can the difference be greater than $4,000$?
(d) Can the difference be equal to $4,000$?

Problem 11.3. The value of the arithmetic expression

$$1,053 - 482$$

is the solution to the unknown addend problem:

$$482 + \boxed{?} = 1,053.$$

Solve this problem using vertical addition with an unknown:

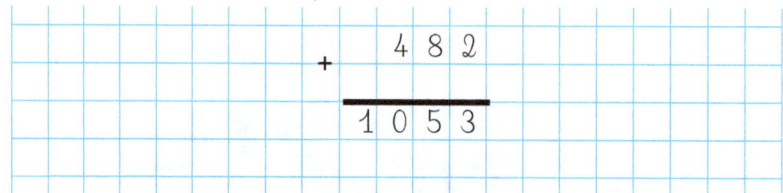

Develop an effective strategy for evaluating subtractions this way.

Problem 11.4. Suppose we evaluate a difference that does not require regrouping. Is it necessarily true that the summation with which we validate the result does not require regrouping? Justify your answer.

Pedagogical problems

Problem 11.5. Teacher Marius asked his class to evaluate the expression:
$$824 - 469.$$
Gili proposed the following solution:
$$824 - 469$$
$$825 - 470$$
$$855 - 500 = \boxed{355}.$$

(a) Explain Gili's strategy.
(b) Is this strategy always applicable?

Chapter 12

Give Me Five!

In this chapter, which concludes the arithmetic part of this volume, we are fantasizing a civilization that invented a different numeral system—a **quinary numeral system**, which is a numeral system based on groupings of five. Such a numeral system is also called a base-5 system. Once we introduce this numeral system, we use it to do arithmetic. Much of the content of the previous chapters is revisited in the context of this alternative numeral system.

You may ask yourself what the role of such extra-curricular material is in this book. One of the challenges educators face in teaching number concepts and arithmetic is to understand the difficulties experienced by children. The subject matter, especially in early school years, is considered to be so elementary by most adults that it is difficult for them to fully empathize with the confusion children experience. A useful technique in training educators is **estrangement**: moving oneself from a familiar place. When adults get acquainted with quinary arithmetic, they are usually confused, and uncertain as to how to exploit their acquaintance with the decimal system. In other words, they feel like children that are faced with base-10 arithmetic for the first time. This is an exceptional opportunity for educators to undergo a learning process similar to their students, and as a result, improve their pedagogical approach.

As a final note, I wish to say that I have taught parts of this chapter on several occasions in grades three to five. I started my lessons by describing an alien civilization where creatures have only five fingers, and as a result develop a base-5 numeral system. I was always amazed by their eagerness to become proficient in this imaginary mathematical setting.

12.1 Quinary numeral system

12.1.1 The shepherd's tale revisited

Recall the shepherd's tale in Chapter 2. Imagine that the wheels of history had spun differently. That shepherd who wants to count the sheep in his herd can only count to five (because this is the number of fingers in one hand). Every time he counts five sheep, he drops a pebble in his jar, where each pebble represents five sheep. When he is almost finished counting, he realizes that he has four sheep he hasn't yet counted. He adds four twigs to his jar, one for each sheep.

Once home, the shepherd empties the content of his jar and finds out that it contains four twigs and a pile of pebbles. This pile comprises more than five pebbles, which means that he can't count them. The resourceful shepherd decides to count those pebbles in the same way as he counted the sheep. He replaces every five pebbles with an arrowhead, until he has less than five pebbles left. At the end of this process, he has three arrowheads, two pebbles and four twigs, as depicted below:

The shepherd knows that each twig represents a sheep, that each pebble represents five sheep, and that each arrowhead represents a combination of five sets of five sheep.

12.1.2 A new numeral system

Next, we describe the numeral system that would result from the recursive bundling of groups of five objects. Such numeral system is called a **quinary system**, which derives from the Latin word for five—*quinque*. Please note the many similarities between the construction of the quinary system and the construction of the decimal system described in Chapters 2 and 9.

In a quinary system, there are distinct words for natural numbers between zero and four. Let's continue using the vocabulary we are accustomed to—the number sequence starting:

> Zero, one, two, three, four.

The number following four represents five items. In a quinary system, we group together five items and represent this bundled five by a new unit. We could call this unit a five (in analogy to a ten in the decimal system), but in order to emphasize the difference between five ones and a bundled five, we will call the latter a **pentad**, derived from the ancient Greek word for five—*penta*. Thus, *pentad* is the number that succeeds *four*. It is a **quinary unit** defined by the conversion rate:

$$\text{one pentad} = \text{five ones}.$$

A pentad is represented by a pebble in the revised version of the shepherd's tale.

We continue the number sequence starting from a pentad:

A pentad and one, a pentad and two, a pentad and three, a pentad and four.

A *pentad and four* represents one bundle of five objects plus four scattered objects. If we add one more object, we again have five scattered objects, which we can bundle together, leaving us with two pentads. The number sequence then continues:

Two pentads, two pentads and one, two pentads and two, two pentads and three, two pentads and four, three pentads,

and further on:

Three pentads and one, three pentads and two, three pentads and three, three pentads and four, four pentads,

and further on:

..., four pentads and three, four pentads and four.

How do we proceed from here? *Four pentads and four* represent four bundles of five objects each plus four scattered objects. If we add one object we obtain five scattered objects, which we can bundle together. We then have five bundles of five objects. In a quinary system, we bundle together these five bundles, and represent this bundle of bundles by a new quinary unit. We will call this new unit an **icosipentad**, that is,

$$\text{one icosipentad} = \text{five pentads}.$$

An icosipentad is represented by the arrowhead in the shepherd's tale (it is a made-up word derived from ancient Greek number prefixes).

With that, we may continue the number sequence until we get to

Four icosipentads, four pentads and four.

The next number is a new quinary unit that represents a bundle of five icosipentads. We will name it an **hectoicosipentad**, that is

$$\text{one hectoicosipentad} = \text{five icosipentads}.$$

(Hectoicosipentad may sound somewhat daunting, but I am not sure that it is more so than quadrillion and quintillion...). Please note that in the revised version of the shepherd's tale, words such as six, seven, ten, thirteen and hundred are nonexistent.

12.1.3 Place-value notation

Now that we have invented a new numeral system based on groupings of five, we will construct a corresponding notational system. The idea is to imitate the notational system we use in our own decimal system, including the use of digits and the place-value principle.

The decimal notation system requires ten digits to represent the whole numbers between zero and nine. The quinary notational system requires just five digits to represent the whole numbers between zero and four. We will stick to our Hindu-Arabic digits:

$$0, 1, 2, 3, 4.$$

In a quinary system that uses place-value notation, we write whole numbers as a sequence of digits. The right-most digit count ones, the digit to its left counts pentads, the one to its left counts icosipentads and the one to its left counts hectoicosipentads (we did not invent names for greater quinary units). Thus, a pentad, which is the number after 4 is denoted by 10, which represents 0 ones, and 1 pentad. Its successor, a pentad and one, is denoted by 11. Thus, the next five elements in the number sequence starting from a pentad are:

$$10, 11, 12, 13, 14.$$

The number after a *pentad and four* is *two pentads*, which we denote by 20. The next several elements of the number sequence are:

$$20, 21, 22, 23, 24, 30, 31, 32, 33, 34, 40, 41, 42, 43, 44.$$

For example, the number whose quinary notation is 43, i.e., *four pentads and three*, represents the following concrete amount:

The number after 44 is an *icosipentad*. Its numeral representation is 100, which stands for 1 icosipentad, 0 pentads and 0 ones. The number sequence starting from 100 proceeds as follows:

$$100, 101, 102, 103, 104, 110, 111, \ldots, 143, 144$$
$$200, 201, 202, 203, 204, 210, 211, \ldots 243, 244$$
$$300, 301, 302, 303, 304, 310, 311, \ldots, 343, 344$$
$$400, 401, 402, 403, 404, 410, 411, \ldots, 443, 444.$$

The number that succeeds 444 is an *hectoicosipentad*. Its numeral representation is 1000, which stands for 1 hectoicosipentad, 0 icosipentads, 0 pentads and 0 ones.

Consider, for example, the number 1 hectoicosipentad, 3 icosipentads, 4 pentads, and 2. Its numeral representation is 1342. Do you have any feel for how large this number is? Can you estimate, for example, how long it will take you to prepare that number of pancakes? An hour? A day? A week? This example shows how much our number sense hinges on intuition acquired over time.

12.2 Addition

Now we will study the evaluation of addition within the quinary numeral system. There is one very important point that we should stress: The interpretations and properties of all four arithmetic properties are independent of the numeral system. Addition represents the joining together of sets of items, and multiplication satisfies the commutative property, whether we use a unary, a decimal or a quinary numeral system. The numeral system only affects representation—not meaning.

Suppose we were learning arithmetic in a civilization that uses a quinary numeral system. Kindergartners and first graders would learn how to add single-digit numbers, first in cases where the sum is also a single-digit number. Note that there are only five single-digit numbers, therefore the number of such sums is significantly smaller than in our system. In fact, since the representation of numbers up to four is the same in both decimal and quinary systems, those **addition facts** seem identical to the ones we are used to. For example:

$$1 + 0 = 1 \qquad 1 + 1 = 2 \qquad 2 + 2 = 4 \qquad 3 + 1 = 4.$$

Things start to look different once we cross the range of a pentad. Consider the following expression:

$$3 + 4.$$

How can we evaluate it? (You will probably find it very difficult not to say seven.) One possible way for evaluating this expression is to apply the *if you can count, then you can add* strategy (see p. 41): Create two sets, one that comprises 3 objects and one that comprises 4 objects. Join the two sets together and count (using the quinary system) the items in the compound set. If you do so, you will discover that the compound set comprises a pentad and two objects, namely,

$$3 + 4 = \boxed{12}.$$

Another strategy for evaluating $3 + 4$ is *adding-to-pentad*. This strategy is illustrated below using manipulatives:

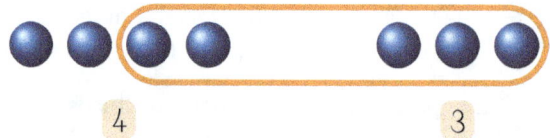

Two items are transferred from the group of four to the group of three. As a result, we have a group that comprises two objects and a group that comprises a pentad of objects. The sum is a pentad and two. This approach relies of the associative property of addition, as follows:

$$3 + 4 = (3 + 2) + 2 = 10 + 2 = \boxed{12}.$$

Similarly, we may find that:

$$4 + 1 = 10 \quad \text{and} \quad 3 + 3 = 11.$$

In fact, every first grader in this imaginary civilization would have to learn the addition facts, which are summarized in the following addition table:

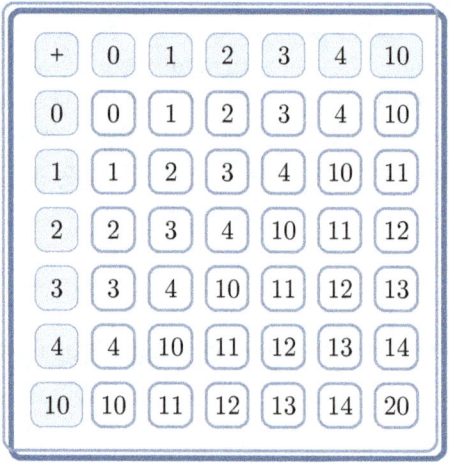

12.2.1 Addition of multi-digit numbers without regrouping

Consider the following expression:

$$3124 + 1210.$$

What is its value? Or in other words:

> What is the result of joining together 3 hectoicosipentads, 1 icosipentad, 2 pentads and 4 ones with 1 hectoicosipentas, 2 icosipentads, 1 pentad and 0 ones?

By now, there is certainly no need for an intermediate step where we replace an icosipentad by an apple, and so on. We simply add each quinary unit separately, and obtain

$$3124 + 1210 = \boxed{4334}.$$

In vertical notation:

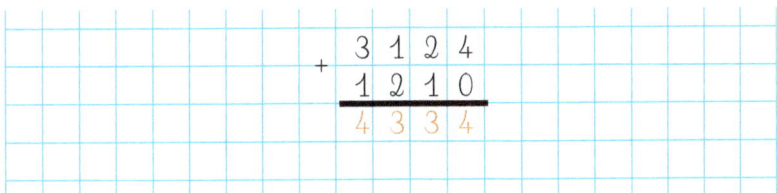

Please note that you can't tell whether the vertical addition shown above uses a quinary or a decimal numeral system. This is so because the digits in each column always sum up to less than five. The distinction between numeral systems is apparent only when regrouping is needed.

12.2.2 Addition of multi-digit numbers with regrouping

Consider the following expression:

$$1342 + 1223.$$

This time we ask:

> What is the result of joining together 1 hectoicosipentad, 3 icosipentads, 4 pentads and 2 ones with 1 hectoicosipentad, 2 icosipentads, 2 pentads and 3 ones?

Adding up each quinary unit separately, we obtain 2 hectoicosipentads, 10 icosipentads, 11 pentads and 10 ones. We can write this sum using a non-standard quinary representation:

$$\boxed{2}_H \; \boxed{10}_I \; \boxed{11}_P \; \boxed{10}_O$$

To obtain a standard representation we need to regroup. First, compose a pentad from five ones, next compose an icosipentad from five pentads, and finally compose an hectoicosipentad from five icosipentads. This sequence of regroupings takes the following form:

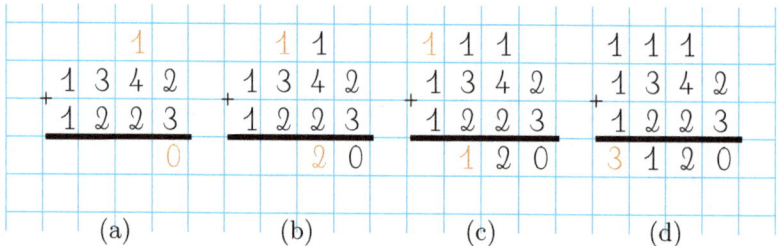

Thus,
$$1342 + 1223 = \boxed{3120}.$$

The corresponding vertical addition algorithm looks as follows:

(a)   (b)   (c)   (d)

12.3 Subtraction

The choice of a numeral system does not affect the interpretations of subtraction, nor the fact that a difference can always be interpreted as the solution to an unknown-addend equation.

Consider the following arithmetic expression:
$$23 - 4.$$
Its value can be interpreted as the number of objects left after removing 4 objects from a set of 23 (2 pentads and 3) objects. This number is the solution to either following unknown-addend equations:
$$\boxed{?} + 4 = 23 \quad \text{or} \quad 4 + \boxed{?} = 23.$$
Since we already know how to evaluate sums in quinary arithmetic, we may check explicitly that the answer is 14.

More subtraction problems are found in the Mathematical problems section.

12.4 Parity

In Chapter 6, we defined even and odd numbers. The two alternative definitions do not depend on the chosen numeral system. In fact, it is possible to determine whether the number of objects in a set is even or odd without any numeral system at all. All we need to do is to try partitioning the objects into two sets that match exactly or equivalently, to try partitioning the objects into pairs.

Is 4 an even number in a quinary system? Yes, because $4 = 2 + 2$. What about 11? Our instincts tell us that 11 must be odd, however this is not true. We are misled if we think of 11 as eleven (a non-existent word in our imaginary world). 11 denotes a pentad and one, and it is even either because
$$11 = 3 + 3,$$
or because
$$11 = 2 + 2 + 2.$$
Likewise, 12, which is the successor of an even number, is odd. We have just discovered that the rule whereby the parity of a number only depends on the parity of its ones digit is not valid in all numeral systems.

Next consider the number 4123. Is it even or odd? We can answer this question as follows (cf. Chapter 6, p. 74):

(a) 1 is an odd number.
(b) A pentad is an odd number, because $10 = 2 + 2 + 1$.
(c) An icosipentad is an odd number, because $100 = 10 \times 10$, i.e. it is the product of two odd numbers.
(d) An hectoicosipentad is an odd number, because $1000 = 10 \times 100$, i.e. it is the product of two odd numbers.
(e) The given number comprises 4 hectoicosipentads, which are even, 1 icosipentad, which is odd, 2 pentads, which are even, and 3 ones, which are odds.
(f) All in all, the given number is the sum of 4 numbers, two of which are odd and two of which are even. By parity arithmetic 4123 is even.

12.5 Multiplication

Multiplication can be interpreted as repeated addition irrespectively of the chosen numeral system. Thus, for example:
$$4 \times 4 = \underbrace{\underbrace{4 + 4}_{13} + 4 + 4}_{22} = \boxed{31}.$$
Products can be evaluated using a variety of strategies, such as those presented in Chapter 7. For example, we can evaluate 4×4 with the help of a number line diagram:

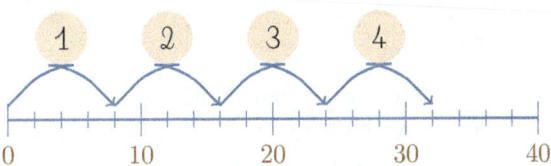

We can also use the associative law. For example:

$$4 \times 3 = 2 \times (2 \times 3) = 2 \times 11 = 22.$$

Eventually, third or fourth graders would be required to be competent at evaluating products with factors in the range of a pentad. These products are summarized in the following multiplication table:

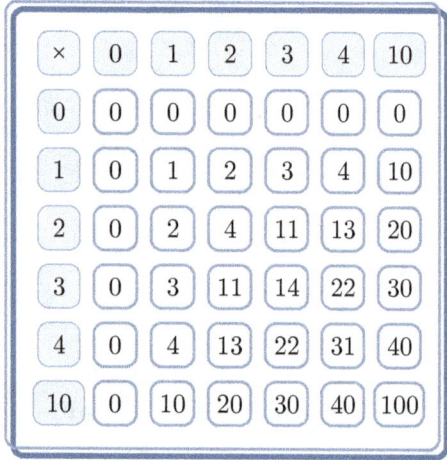

12.6 Division

The choice of a numeral system does not affect the interpretations of division, nor the fact that a quotient can always be interpreted as the solution to an unknown-factor equation.

Consider the following arithmetic expression:

$$13 \div 2.$$

Its value can be interpreted as the size of a portion when 13 (a pentad and three) objects are partitioned into two sets of equal size. This number is the solution to either of the following unknown-factor equations:

$$\boxed{?} \times 2 = 13 \quad \text{or} \quad 2 \times \boxed{?} = 13.$$

Since we already know how to evaluate products in quinary arithmetic, we may check explicitly that the answer is 4.

There are various ways to obtain this solution. One possibility is to find the number in the multiplication table that yields 13 when multiplied by 2. Another strategy described in Chapter 8 is forward counting, as illustrated below:

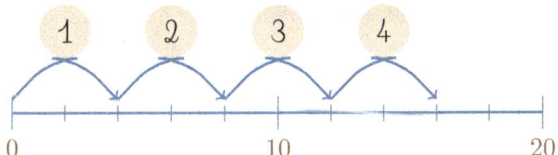

Mathematical problems

Problem 12.1. Match each set of objects with an appropriate quinary numeral:

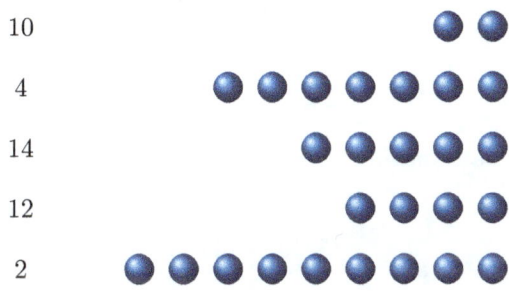

Problem 12.2. Evaluate the following sums (within the range of two pentads):

$$3 + 2 \qquad 4 + 4$$
$$0 + 3 \qquad 3 + 3$$
$$4 + 2 \qquad 12 + 3$$
$$2 + 11 \qquad 3 + 4$$
$$10 + 3 \qquad 10 + 0$$

Problem 12.3. Evaluate the following sums (within the range of an icosipentad):

$$14 + 11 \qquad 32 + 4$$
$$23 + 10 \qquad 4 + 22$$
$$13 + 13 \qquad 41 + 4$$

Problem 12.4. Evaluate the following sums using vertical addition:

$$444 + 123$$
$$34102 + 14334.$$

Problem 12.5. Complete the following sequences:

$$2, 4, 11, 13, \underline{\ \ }, \underline{\ \ }, \underline{\ \ }, \underline{\ \ }$$
$$3, 11, 14, \underline{\ \ }, \underline{\ \ }, \underline{\ \ }, \underline{\ \ }$$
$$4, 12, 20, \underline{\ \ }, \underline{\ \ }, \underline{\ \ }, \underline{\ \ }$$

Problem 12.6. Evaluate the following differences (minuends within the range of two pentads):

$$14 - 3 \qquad 14 - 11$$
$$11 - 3 \qquad 10 - 0$$
$$12 - 4 \qquad 4 - 2$$
$$10 - 2 \qquad 13 - 4$$

Problem 12.7. Evaluate the following differences (minuends within the range of an icosipentad):

$$31 - 4 \qquad 100 - 13$$
$$21 - 14 \qquad 100 - 44$$
$$40 - 11 \qquad 23 - 14$$

Problem 12.8. Evaluate the following differences using vertical subtraction:

$$1000 - 123$$
$$34102 - 14334.$$

Problem 12.9. Which of the following numbers are even?

$$4, 12, 43, 142, 44444.$$

Propose a general method for determining the parity of a number.

Problem 12.10. Does multiplication by 10 (i.e., by a pentad) result, as in a decimal system, in each digit moving one column to the left and placing 0 in the ones column? Justify your answer.

Problem 12.11. Evaluate the following product (factors within the range of one pentad):

$$3 \times 2 \qquad 4 \times 4$$
$$3 \times 3 \qquad 4 \times 10$$

Problem 12.12. Evaluate the following products (factors within the range of an icosipentad):

$$22 \times 2 \qquad 22 \times 12$$
$$13 \times 44 \qquad 22 \times 22$$

Problem 12.13. Evaluate the following quotients:

$$13 \div 2 \qquad 14 \div 3$$
$$22 \div 3 \qquad 13 \div 4$$

Chapter 13

Introduction to Geometry

Geometry is a branch of mathematics that studies shapes. Historically, geometry was invented to respond to people's natural needs, such as the measurement of length, area, and volume; the setting of borders, navigation, architecture and more. Like arithmetic, geometry began as a mathematical discipline that models concrete objects. It took generations of geometers until this discipline emerged as a theory that permits abstract reasoning about shapes and magnitudes. This chapter presents a general introduction to geometry, and advances some of the concepts and terminology used in subsequent chapters.

13.1 Euclidean geometry

Geometry evolved independently in several ancient cultures, notably in Egypt, Babylon and Greece. Two early Greek mathematicians whose work is known to schoolchildren are Thales of Miletus (635–543 BC), recognized for having developed an ingenious way for calculating the height of a pyramid, and Pythagoras (582–496 BC), remembered for the theorem that bears his name.

The founding father of geometry is the Greek mathematician Euclid (325–265 BC). Euclid was the one who transformed geometry from a plethora of vague ideas and computational practices into a formal mathematical theory. He transformed geometry into an **axiomatic**, or **deductive** theory. A deductive theory starts by putting forward a set of **fundamental concepts**, i.e., concepts that are not defined in terms of former concepts. Euclid's chief geometric concepts are **space**, **planes** and **lines**. In addition, a deductive theory announces certain **postulates**, or **axioms**. Just as fundamental concepts are indefinable, postulates are neither provable nor refutable; they are taken for granted as absolute truth. For example, one of geometry's postulates is that through every pair of points passes a unique straight line.

Equipped with fundamental concepts and axioms, deductive geometry defines new concepts and discovers new facts satisfied by those concepts. Every new concept has to be defined rigorously by means of existing concepts, and every new fact is accepted as a truth only if it is justified rigorously based on other facts whose truth has already been established. The assertion that something is true is called a **theorem**. The sequence of logical arguments justifying this assertion is called a **proof**. This requirement of mathematical rigor reflects the Greeks' belief that inference based on logical reasoning is superior to inference based on sensory input (i.e., do not believe what you see).

13.2   Elementary school geometry

In most countries, deductive geometry is not studied before middle school, and often even later. In elementary school, geometry is similar to what it was in pre-Euclidean times—informal, descriptive and application-based. Specifically, elementary school geometry includes the following constituents:

(a) Basic geometric concepts (e.g. points, lines, angles and parallelism).
(b) Geometric figures (e.g. triangles, circles and pyramids) and their defining properties.
(c) Properties of geometric figures (e.g. the opposite sides of a rectangle are parallel and have equal length).
(d) Set-theoretic relations satisfied by classes of geometric figures (e.g., every square is a rectangle).
(e) Measurements (e.g. lengths, angles and areas).
(f) Pre-deductive reasoning: explanation of geometric facts using informal language and manipulatives.

As in other fields of mathematics, the study of geometry requires a gradual transition from the concrete to the abstract. Elementary school geometry provides children with a hands-on experience with geometric concepts through the manipulation of solid objects and illustrations.

13.3 Set-theoretic concepts

Set theory is the mathematical branch that studies sets. Set-theoretic concepts play a central role in geometry (in fact, throughout the entirety of mathematics) for two main reasons. First, geometric figures, which are the subject matter of geometry, are sets of points. Second, polygons, circles, and pyramids, just to name a few, are sets of geometric figures (i.e., sets of sets of points). No wonder the study of geometry involves an extensive use of set-theoretic concepts and terminology.

Common set-theoretic concepts that are used in subsequent chapters are listed below:

(a) A **set** is a collection of objects. The objects contained in a set are called **elements**.
(b) An object is said to **be in a set**, or **belong to a set** if it is one of its elements. Reciprocally, a set **includes** an element if the element belongs to it.
(c) A set that includes no elements is called an **empty set**.
(d) Set A is **contained** in Set B (or Set B contains Set A) if every point that belongs to Set A belongs also to Set B.
(e) Set A is a **subset** of Set B if it is contained in Set B.
(f) The **union** of Sets A and B is a set that includes all the elements of Set A and all the elements of Set B, and only those elements.
(g) The **intersection** of Sets A and B is a set that includes only those elements that belong both to Set A and Set B.
(h) Two sets are said to be **disjoint** if there exists no element that belongs to both.
(i) Two sets are said to be **equal**, or to **coincide** if every elements that belongs to one set also belongs to the other set, and vice versa.
(j) Two sets are said to be **different** if they are not equal.

13.4 Three-dimensional space

Since the times of Euclid, a geometric theory begins by postulating the existence of certain **basic concepts**, or **basic entities**. As geometry was created originally as a model of the universe, those basic entities are all abstractions of concrete entities from our environment. The geometric entity **space** is an abstraction of the entire universe, at least as it was envisioned in ancient Greece. The universe was perceived as comprising of atomic, dimensionless positions. The abstract equivalent of a position in the universe is called a **point**. Thus, in its most primitive sense, space is a collection of points, or in other words, a set whose elements we call points.

The set of points that form space is endowed with a structure, which reflects our perception of the universe. There is a notion of **neighborhoods** of points, which are subsets of space that surround that point. Every two points can be joined smoothly by a **path**. Space is **three-dimensional** in the following sense: all the directions that emanate from a given point are spanned by a combination of three axes. This is consistent with our experience that we can reach any point by a combination of displacements along a forward-backward axis, an up-down axis, and a right-left axis.

Even though the mathematical notion of space is thousands of years old, it was not until the 20th century that space was defined rigorously. The formal definition of space is beyond the scope of this book. Yet, we can manage perfectly without a formal definition, as did generations of geometers since Euclidean times.

13.5 Geometric figures

Space represents the entire universe. We are rarely interested in the entire universe. We are usually interested in parts of the universe, such as the region occupied by a galaxy, the surface of the earth, the line that connects Cincinnati to Denver, or the interior of a water tank. The mathematical abstraction for a part of the universe is a part of space. Since space is a set of points, a part of space is a subset of all the points that form space. A subset of space is called a **geometric figure**. A triangle, a sphere, and a prism are all instances of geometric figures. Please note that a geometric figure can be any subset of space, regardless of whether there is a designated name that classifies it.

When we first introduced the concept of a number in Chapter 1, we emphasized the importance of a gradual transition from the concrete to the abstract. The same is true in geometry. Children (and adults) need to visualize geometric figures in order to understand what they are and what properties they satisfy. Visualizations can be of several sorts. Bodies, which form a category of geometric figures, are best visualized by solid objects. Dedicated computer programs offer a variety of tools to visualize geometric objects. The most common visualization tool is drawings, and of course they are used in books.

It is important to note that geometric figures are idealized objects, unlike the objects used to visualize them. Consider a point, for example. A true point is dimensionless. When we illustrate a point we touch the paper with the tip of our pencil or pen. The mark left on the paper is not dimensionless. It is a speck of graphite or ink, which viewed through a microscope, exhibits a rich structure that has nothing to do with the mathematical notion of a point. Visualizations should never be expected to be identical to the objects they represent; they are just leads to our imagination.

A single point, being a subset of space, is a geometric figure, albeit not a very interesting one. We normally study and visualize geometric figures that contain many points (usually infinitely many). Given a visualization of a geometric figure, it is often useful to label certain points in that visualization (e.g., the vertices of a triangle). Assigning names to points helps us to formulate unambiguous statement about those geometric figures. Each point should be labeled differently in order to avoid ambiguity. Like many other mathematical objects, it is conventional to label points by letters, the standard choice being uppercase letters as depicted below:

13.6 Congruence

A central theme in mathematics is the concept of **sameness**. Whenever a new class of mathematical objects is introduced, one always defines a notion of sameness between elements of that class—a criterion for two elements of that class to be considered as same. We have already encountered two instances of sameness in arithmetic:

(a) **Equivalence between sets**: two sets are considered as equivalent if there exists between them an exact matching (Chapter 1, p. 11).

(b) **Equivalence between arithmetic expressions**: two arithmetic expressions are equal if they have the same numerical value.

A concept of sameness is also introduced for geometric figures. Since a geometric figure is a subset of space, two geometric figures are equal if they are the same subset of space, or in simpler terms, if they occupy the same location. This notion of sameness is too restrictive. Consider the two geometric figures illustrated below:

The two figures do not coincide in space, and yet, we are inclined to say that they are the same; one figure can be obtained from the other by translating it and then rotating it, *"without changing its shape"*. Euclidean geometry postulates a notion of sameness between geometric figures that is not dependent on position or orientation. This notion of sameness is called **congruence**; the above geometric figures are congruent.

Euclidean geometry introduces congruence as a basic concept. Intuitively, two figures are congruent if one can be obtained from the other through a combination of so-called Euclidean transformations: translations, rotations, and reflections. Euclidean transformations, and consequently also congruence, were formally defined only with the introduction of analytic geometric in the 17th century. Yet, as with many other geometric concepts, our intuitive perception of congruence does not require a formal definition. Children understand the notion of two figures being congruent very well: if we place one figure on top of the other, they will match exactly.

Congruence, like all other types of sameness, is an equivalence relation. This means that it satisfies three properties:

(1) **Reflexivity**: every geometric figure is congruent to itself.
(2) **Symmetry**: if Figure A is congruent to Figure B, then Figure B is also congruent to Figure A.
(3) **Transitivity**: if Figure A is congruent to Figure B, and Figure B is congruent to Figure C, then Figure A is also congruent to Figure C.

Congruence is denoted by the symbol $\cong$, which is reminiscent of the equality sign.

13.7 Measurements

Measurement is the assignment of a numerical value to an attribute of an object. Attributes of objects to which we can assign a numerical value are called **measurable attributes**. Examples of measurable attributes learned in school are **length** (e.g., of a rod), **area** (e.g., of a field), **volume** (e.g., of a water tank), **angle** (e.g., between two rods), **mass** (e.g., of a watermelon), and **temperature** (e.g., of boiling water). Part of the study of geometry is concerned with the measurement of measurable attributes of geometric objects.

All measurable attributes share two traits:

(1) **Comparability**: Given two objects (e.g., two pencils) endowed with the same measurable attribute (e.g., length), it is possible to determine whether both objects have equal amounts of that attribute, or whether one object has more of that attribute than the other (e.g., it is longer). The ability to compare two objects with respect to a certain attribute induces an order relation between objects that have this attribute. For example, when the objects are pencils and the attribute is length, we can sort a stack of pencils from shortest to longest.

(2) **Measurability**: Given two objects (e.g., a playing die and a pack of flour) endowed with the same measurable attribute (e.g., mass), it is possible to determine how many times the measurable attribute of one object is contained in the measurable attribute of the other object. For example, it is possible to determine how many times the mass of the die is contained in the mass of the pack of flour. Differently stated, measurable attributes allow for quotative

division. *Measurement results are quotients of quotative division*, where the dividend is the object we measure and the divisor is a standard object, which we call a **measuring unit**.

Please note that there is no point in asking what the meaning of a measurable attribute is. A measurable attribute is defined by how we compare it and by how we measure it. *I can't tell you what area is. I can only tell you how to measure it.*

In this volume, we study the measurable attributes length (Chapter 15), angle (Chapter 16) and area (Chapter 18).

Chapter 14

Planes and Lines

In Chapter 13, we presented the principles of Euclidean geometry as an axiomatic theory, which rests on a selected set of basic concepts and postulates. In Euclidean geometry, the entire universe is modeled by an entity called **space**, whose elements are called **points**. **Geometric figures**, which are the subject matter of geometry, are sets of points. In this chapter, we introduce two families of geometric figures that are central to Euclidean geometry: **planes** and **lines**.

14.1 Planes

Ancient Greek geometers identified certain subsets of space, which they called **surfaces**. A surface is a **two-dimensional** geometric figure. That is, all the directions that emanate from a point are spanned by combining motions along two axes (e.g., a forward-backward axis and a right-left axis). Surfaces are abstractions of real entities, such as the surface of the earth, or the sail of a boat.

Among all surfaces, the Greeks postulated the existence of a distinguished sub-category—**planes**. Intuitively, a plane is a flat surface that stretches indefinitely in all directions. This characterization is not very informative, unless we specify explicitly what we mean by flatness and by extension to infinity. Like the concept of space, the concept of a plane had to wait over two thousand years from Euclidean times until it was rigorously defined. In this book as well, we will make do without formal definitions. Beautiful and useful geometry can be developed by perceiving a plane in the same way as did Greek geometers: as the surface of a smooth floor that extends indefinitely in all directions.

The most common visualization of a finite portion of a plane is the surface of a sheet of paper. Sometimes, we visualize a part of a plane as embedded within three-dimensional space. A typical illustration of a plane is depicted below:

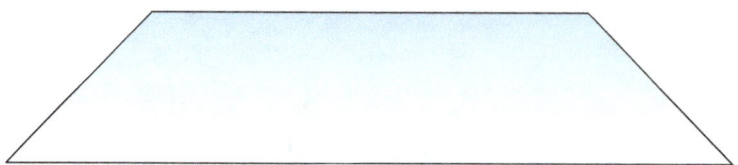

14.1.1 Plane geometry

Greek geometers viewed planes as idealizations of the surface of Earth (long before it was realized that Earth is in fact a sphere). Many practical applications of geometry, such as the setting of borders and navigation, are concerned with entities that lie on the surface of Earth. The geometric figures that are used in the mathematical modeling of those applications are all subsets of a plane. A geometric figure that is a subset of a plane is called a **planar figure**. The geometric theory that studies figures in a plane is called **plane geometry**. A large fraction of the school curriculum is devoted to plane geometry.

14.2 Lines

Ancient Greek geometers identified certain subsets of space, which they called **curves**. A curve is a **one-dimensional** geometric figure. That is, all the directions that emanate from a point are along a single axis (backward/forward). Curves are abstractions of real entities, such as the border of a country, or the trajectory of a projectile.

Among all curves, Greek geometers postulated the existence of a distinguished subcategory **lines**. Intuitively, a line is a straight curve that stretches to infinity in both directions. Exactly as the plane is a subset of space that encompasses the notion of flatness, so is the line, a subset of a plane that encompasses the notion of straightness. Like the flatness of a plane, the straightness of a line will be left without a formal definition. Children intuitively know what lines are.

We illustrate "geometric lines" by outlining "concrete lines" with a ruler. Of course, we can only draw a finite segment of a line, as the one below:

14.3 Postulates and theorem

14.3.1 Three Euclidean postulates

Thus far, we have postulated the existence of space, which is a collection of points, and the existence of planes and lines. In order to define new geometric figures and establish their properties, we must make some assumptions about the nature of our basic entities.

It is useless to postulate the existence of planes and lines if we do not endow them with some properties. This is precisely where Euclidean postulates come into play. Euclidean postulates endow planes and lines with properties that reflect the Greeks perception of what planes and lines are.

The first Euclidean postulate is an abstraction of something mundane: once we fix two points, no matter which two points, there is a unique straight line that passes through those points.

> **Postulate 14.1.** For every pair of points there exists a unique line that passes through both points.

Postulate 14.1 is illustrated below. Points A and B are arbitrarily chosen points. The line in the figure contains both points, or differently stated, passes through both points. Postulate 14.1 can be broken into two separate statements:

(a) There exists a line that passes through both points.
(b) Given such a line, there is no other line that passes through both points.

Now suppose that there are three points, A, B and C. What can we infer from Postulate 14.1?

Since Postulate 14.1 concerns pairs of points, and since we can form three different pairs from every three points, three conclusions follow from Postulate 14.1:

(a) There exists a unique line that passes through Points A and B.
(b) There exists a unique line that passes through Points A and C.
(c) There exists a unique line that passes through Points B and C.

A priori, these three lines may be the same line, or they could be different lines. Postulate 14.1 says nothing about such eventualities.

Consider the line that passes through Points A and B. There are two possible scenarios regarding the location of Point C relative to this line:

(a) Point C lies on the line that passes through Points A and B:

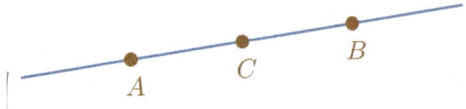

(b) Point C does not lie on the line that passes through Points A and B:

The next postulate applies to the second scenario:

Postulate 14.2. Let A, B and C be three points that do not lie on a line (i.e., there is no line that passes through all three points). There exists a unique plane that includes all three points (see figure below).

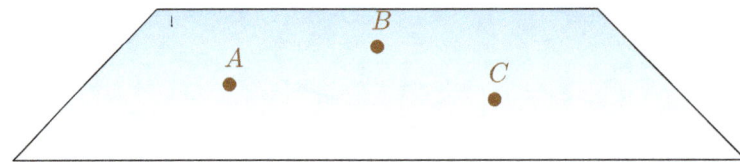

Postulate 14.2 is similar in spirit to Postulate 14.1. Postulate 14.1 states that every two points determine a unique line. Postulate 14.2 states that every three points determine a unique plane, provided that they are not aligned.

Postulate 14.2 has a practical aspect: a three-legged stool standing on a flat surface is always stable. Think of the bases of the legs of the stool as three points and think of the floor as part of a plane. Postulate 14.2 implies that the three bases of the legs of the stool can touch the floor simultaneously, even if the legs are of uneven length and the floor is tilted. This is also the reason why photographers use a three-legged tripod to balance their camera.

The next postulate formalizes the following fact: if we take a straight rod and bind it to a flat surface at two points, then the entire rod is bound to that surface.

Postulate 14.3. Let A and B be two points on a given plane. Then, the line that passes through A and B is contained in that plane.

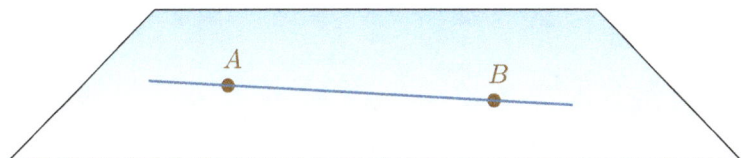

14.3.2 A sample theorem

Euclidean postulates are a set of presumptions about certain geometric figures. For a collection of postulates to form the basis of a useful geometric theory, it has to provide a sufficiently broad characterization of basic geometric concepts. Postulates 14.1–14.3 provide us with notions of inclusion relations between space, planes, lines and points. These postulates do not provide us with a sufficiently rich foundation, hence more postulates are to follow.

Yet, even with an arsenal of postulates as minimal as Postulates 14.1–14.3, we can already draw a few conclusions. We now exemplify how a new fact can be inferred from older facts. We formulate a **theorem** (i.e., we assert that a certain fact is true), and then we prove it through deductive reasoning. We will take advantage of this sample theorem-proof to exhibit a very powerful deductive approach, known as **proof by contradiction**.

Theorem 14.1. *Two different straight lines cannot intersect at more than one point. They either intersect once, or they do not intersect at all.*

Proof. The idea behind a proof by contradiction is the following: one first assumes that the theorem is false. Based on this assumption, one deduces that a certain fact known to be false is true. The only way to resolve this contradiction is to conclude that the initial assumption was wrong, namely, that the theorem was true.

In the present case, suppose that the theorem is false. Then, there are two lines, a and b, that intersect more than once. Let A and B be two such points of intersection:

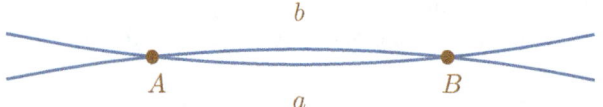

Since Points A and B lie on both Line a and Line b, they are two points through which passes more than one line. This contradicts Postulate 14.1, according to which there exists a unique line that passes through any two given points. Therefore two lines cannot intersect more than once. □

14.3.3 Ordering of points on a line

Postulates 14.1–14.3 provide us with the knowledge that certain geometric objects, planes and lines, exist and are unique. They do not provide us with any notion of relations between points. The next postulate is concerned with the relative position of points that lie on the same line.

Euclidean geometry introduces a concept of **betweenness** (the property of an object being in between two other objects) for points that lie on the same line. Betweenness is a basic concept, which acquires its meaning from the following postulate:

> **Postulate 14.4.** For every two points, A and B, there exists a third point, C, that lies between A and B.
>
> For every two points, A and B, there is a third point, C, such that B lies between A and C:
>
> Given four points, A, B, C, and D, if B lies between A and C, and C lies between B and D, then B (and C as well) lies between A and D:

The first part of this postulate states that lines are **dense**: between any two points there is always at least one other point (and consequently, there are infinitely many points between any two points). The second part states that a line has no end; there are always more points beyond any given point. The third part expresses the transitive property of betweenness.

14.4 Segments and rays

14.4.1 Line segments

The notion of betweenness for points on a line enables us to define one of the most important families of geometric figures—**line segments**, or in short, **segments**. Intuitively, a segment is a part of a line that is delimited by two endpoints. The formal definition is not that different:

> **Definition 14.1.** A segment is a geometric figure that comprises two points and all the points that lie in between them.

The two points that delimit the segment are called the **endpoints** of the segment. The common notation of a segment is by its endpoints. A segment whose endpoints are A and B is denoted by either AB, or by BA. Please note that a segment has no preferred direction, hence the order of the endpoints does not matter.

14.4.2 Rays

Another geometric object that can be defined with the notion of betweenness is a **ray**. Intuitively, a ray is a semi-infinite line.

Given two points, A and B, those two points partition the line that passes through them into three disjoint parts:

(a) The segment AB.
(b) Points that are beyond A, that is, A is between those points and B.
(c) Points that are beyond B, that is, B is between those points and A.

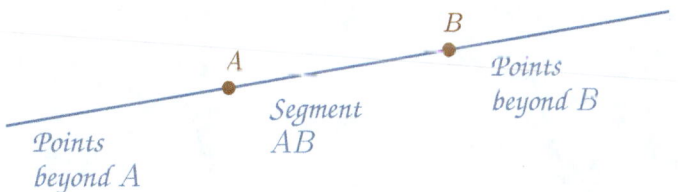

Definition 14.2. Let A and B be two points. The ray **from A through** B consists of the union of the segment AB and those points on the line that passes through Points A and B that are beyond Point B.

An illustration of a ray is shown below; the ray extends indefinitely on one side.

Point A is called the **endpoint**, or the **vertex** of the ray. Please note that the ray from A through B and the ray from B through A are different geometric figures. It is customary to denote a ray by two points—the vertex and some other point on the ray. For example, the ray AB is the ray from A through some other point B.

14.4.3 Broken lines

A **broken line** is a chain of segments that are concatenated end-to-end. The definition is the following:

Definition 14.3. A broken line is a geometric figure that consists of an ordered set of points together with the set of segments that connect each point to its successor (a set of points is called ordered if we endow it with an order relation: first, second, third, etc.).

The points that form the "skeleton" of the broken line are called **vertices**. The segments that connect the vertices are called **edges**. A broken line is called **closed** if it also includes the segment that connects the first vertex to the last vertex. Otherwise, it is called **open** (see figure below).

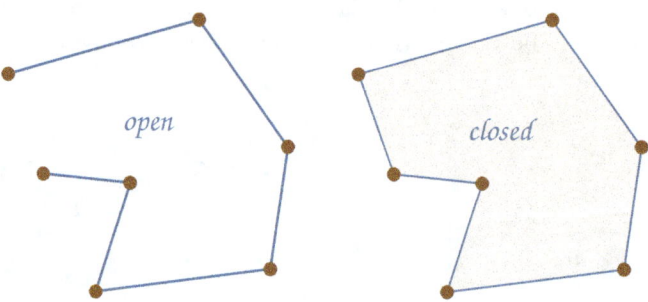

14.5 Segment arithmetic

14.5.1 Segment comparison

In Chapter 1, we defined a notion of order between sets: Set A is said to be smaller than Set B if it can be matched to a subset of Set B. The same principle defines an order relation between segments:

> **Definition 14.4.** Segment a is **shorter** than Segment b (reciprocally, Segment b is **longer** than Segment a) if it is congruent to a subset of Segment b.

We denote the property of Segment AB being shorter than Segment CD using the standard inequality signs, namely, $AB < CD$ or $CD > AB$.

The order relation between segments satisfies the same properties as all order relations, namely:

(1) Every segment is equal to itself, $AB \cong AB$.
(2) Given two segments, AB and CD, exactly one of the following holds:
  (a) $AB \cong CD$.
  (b) $AB < CD$.
  (c) $AB > CD$.
(3) Transitivity: if $AB < CD$ and $CD < EF$, then $AB < EF$.

In the next chapter, the longer/shorter relation will be connected to a measurable attribute of segments—**length**.

14.5.2 Segment addition

Addition, as we learned in Chapter 4, is an operation that takes two numbers, which we called **addends**, and yields a third number, which we called a **sum**. Numbers are not the only mathematical entity for which addition can be defined. We now define an addition of segments.

The addition of segments is very intuitive. Given two segments, their sum is a segment formed by concatenating them, or equivalently, by concatenating segments that are congruent to the given segments.

Suppose we are given two segments, AB and CD:

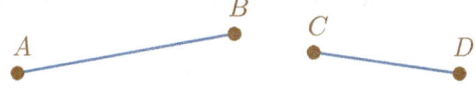

To add them, we construct a ray with vertex E, and mark on that ray a point F, such that $EF \cong AB$:

In other words, we constructed on the ray a "copy" of Segment AB. Then, we allocate on the ray EF a point G, such that $FG \cong CD$:

In other words, we constructed on the ray a "copy" of Segment CD, which is **adjacent** to the copy of Segment AB.

The segment EG is called the sum of the segments AB and CD. We write:
$$EG \cong AB + CD.$$

This addition of segments is very similar to the addition of numbers. Both represent the joining together of two sets. Moreover, both additions satisfy the same properties. For example, segment addition is commutative: for every two segments, AB and CD,
$$AB + CD \cong CD + AB,$$

Likewise, segment addition is associative. For every three segments, AB, CD and EF,
$$(AB + CD) + EF \cong AB + (CD + EF).$$

14.5.3 Segment subtraction

If we can add segments, then we can also subtract them. Recall that subtraction is an operation whose outcome can be interpreted as the solution to an unknown-addend equation. Given two segments, AB and CD, we define their difference to be the segment which, when added to CD, yields a segment that is congruent to AB. That is, the difference
$$AB - CD,$$
is the solution to the unknown-addend equation,
$$CD + \boxed{?} \cong AB.$$

The difference $AB - CD$ can be constructed as follow: Mark on Segment AB a point, E, such that
$$AE \cong CD.$$

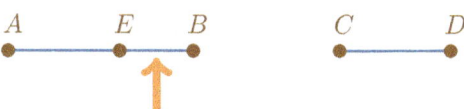

Now, the segment AB is the sum of the segments AE and EB. Since AE is congruent to CD it follows that

$$CD + \boxed{EB} \cong AB.$$

Thus, EB is the addend that solves the unknown-addend equation, or in other words,

$$AB - CD \cong EB.$$

14.5.4 Segment multiplication

If we can add and subtract segments, can we also multiply segments? For example, we can add chairs, but can we multiply chairs? No, a chair times a chair is neither a chair not two chairs; it is meaningless. Likewise, the product of two segments is meaningless. However, since we can add segments, then there is a sense to the **repeated addition** of segments. In other words, we can multiply a segment by a number. Please note that this interpretation of segment multiplication requires the multiplier to be a number and the multiplicand to be a segment. The product is a segment.

The following figure illustrates the product of 3 times a given segment:

14.5.5 Segment division

If we can multiply segments (by numbers) we can also divide them. Recall that division is an operation whose outcome can be interpreted as the solution to an unknown-factor equation. In this context, there is a distinction between partitive division and quotative division.

In **partitive division**, the unknown factor is the multiplicand. We are given the multiplier (i.e., the number of portions) and the product (i.e., the total amount), and the unknown is the multiplicand (i.e., the size of each portion). In the context of segment division, the dividend is a segment, the divisor is a number and the quotient is a segment. Consider for example the expression:

The quotient is a segment that, multiplied by 5, yields the dividend. Written as an unknown-factor equation:

$$5 \times \boxed{?} \cong$$

In **quotative division**, the unknown factor is the multiplier. We are given the multiplicand (i.e., the size of each portion) and the product (i.e., the total amount), and the unknown is the multiplier (i.e., the number of portions). In the context of segment division, the dividend is a segment, the divisor is a segment too and the quotient is a number. Consider for example, the expression

The quotient is the number of times that the divisor is contained in the dividend, or equivalently, the number whose product with the divisor yields the dividend. Written as an unknown-factor equation:

$$\boxed{?} \times \bullet\!\!-\!\!-\!\!-\!\!\bullet \;\cong\; \bullet\!\!-\!\!-\!\!-\!\!-\!\!-\!\!-\!\!\bullet$$

Quotative division of segments is the main foundation for the next chapter, which introduces the concepts of **length** and its measurement.

Chapter 15

Length

In Chapter 13, we introduced the concept of **measurable attributes**. A measurable attribute of an object is a property of that object that can be assigned a numerical value. Not every attribute that can be assigned a numerical value is measurable: for example, a bus may be associated with a route number, but this number wouldn't qualify as a measurable attribute. Measurable attributes must be accompanied with a method for determining how many times that attribute in one object is contained in that attribute in another object. Stated differently, a measurable attribute is endowed with a concept of **quotative division**. **Measurement** is the process by which we find out how many times a measurable attribute in some standard object—a **measuring unit**—is contained in the same measurable attribute in the object under consideration.

This chapter is dedicated to the measurable attribute **length**. Length is an attribute of one-dimensional geometric figures, i.e., **curves** (see the definition in Chapter 14, p. 172). We start by defining length measurement for the simplest one-dimensional figures—**segments**. We then proceed to define the length of more complicated curves.

15.1 Length comparison

An order relation between segments was defined in Chapter 14 (p. 179). Two segments are said to be of **equal length** if they are congruent to each other. Segment a is said to be **shorter** than Segment b if it is congruent to part of it; reciprocally, Segment b is said to be **longer** than Segment a.

Suppose we were given two objects, for example, pencils. Pencils are three-dimensional bodies, but nevertheless, we can assign a length to the segment that connects their tips to their bottoms. The question is how to determine *in practice* which of two objects is longer.

15.1.1 Comparison by juxtaposition

Two segments are congruent if we can place one on top of the other such that they match exactly. Strictly speaking, when we say that two geometric figures are one on top of the other, we mean that they occupy the exact same position in space. Obviously, we can't have two material objects occupy the exact same position. The closest thing we can do is to place them in proximity.

Given two objects, say pencils, we can place them such that their bottom parts are lined up. If their tips are also lined up, then the segments that connect their ends are congruent, and we conclude that the pencils are of equal length. If, however, the tip of Pencil a is lined up with a point that lies between the bottom part and the tip of Pencil b (see figure below), then we deduce that Pencil a is shorter than Pencil b.

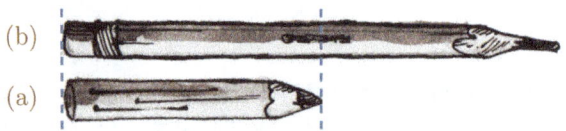

15.1.2 Comparison by transitivity

In many situations, we need to compare the length of objects that cannot be displaced (e.g., roads). In such cases, comparison by juxtaposition is unfeasible. Instead, it is possible to exploit the transitive property of order relations, which in the present case reads as follows:

> If Segment a is equal to, or shorter than Segment b, and Segment b is shorter than Segment c, then Segment a is shorter than Segment c.

Transitivity can be exploited as follows: Given two segments, a and b, we generate a third segment, c, which is congruent to Segment a, but unlike Segment a, can be displaced. For example, Segment c can be sketched on paper, or it can be represented by a thin rod. We then juxtapose Segments b and c.

There are three possible scenarios:

(1) Segment c and Segment b have equal length. In this case, we deduce that Segment a and Segment b have equal length.
(2) Segment c is shorter than Segment b. In this case, we deduce that Segment a is shorter than Segment b.
(3) Segment c is longer than Segment b. In this case, we deduce that Segment a is longer than Segment b.

15.1.3 Comparison by concatenation

Suppose we want to compare the length of two soccer fields (more precisely, the length of their side lines). Obviously, we can't juxtapose them. Theoretically, we could use transitivity by constructing a very long rod that is congruent to the side line of one of the fields. This isn't a practical solution either, and it is totally impractical for even longer objects (e.g., an interstate highway).

A more practical method of comparison is based on the fact that segments can be added (Chapter 14, p. 179). In its most rudimentary version, the comparison method goes as follows:

(a) Take a rod of manageable size, and produce many identical copies of it.
(b) Concatenate those rods, end to end, along the side line of one of the soccer field, until the resulting chain of rods connects its two endpoints.
(c) Count how many rods were used.
(d) Repeat the same procedure for the other soccer field.
(e) The soccer field that requires more rods is longer.

This comparison method is sketched below:

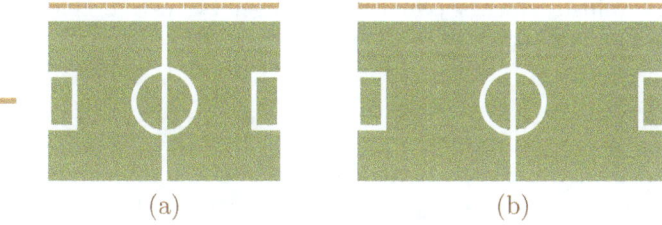

(a)   (b)

A single rod is displayed on the left of the figure. The side line of Field a contains that rod 12 times, whereas the side line of Field b contains that rod 16 times. Thus, we conclude that Field b is longer than Field a.

15.2   The length of a segment

The procedure we have just described, counting how many times a segment is contained in another segment, is precisely what is meant by the measurement of the length of a segment. When we measure the length of a segment, we are always equipped with the following:

(1) A segment whose length we measure.
(2) A segment which we call a **measuring unit**.
(3) A practical means of counting how many times the measuring unit is contained in the segment whose length we measure.

The outcome of the measurement is the number obtained in the last step.

It is customary to assign names to measuring units. Suppose that we called the rod used in the previous figure a *perch* (an ancient Roman length unit). Then we would say that Field a is 12 perches long, whereas Field b is 16 perches long.

15.2.1 Standard measuring units

Suppose we sent two surveyors to measure the lengths of the two soccer fields depicted above, each equipped with a different measuring unit. The first surveyor uses *perches* and the second surveyor uses *cubits* (an ancient Egyptian length unit). The first surveyor reports that Field a is 12 perches long, whereas the second surveyor reports that Field b is 31 cubits long. Which soccer field is longer?

There is no way we can answer this question unless we know the ratio between a perch and a cubit. Had the two surveyors used the same measuring unit, whichever one, we could have determined which soccer field is longer by just comparing the numerical output of their measurements.

This example demonstrates the usefulness of **standard measuring units**. If every individual adopted his or her own measuring units, it would be practically impossible to communicate length measurement data that is mutually comprehensible. In ancient times, when the communication between different civilizations was limited, each civilization used its own standard units. In modern times, especially in view of a rapidly globalizing world, the number of standard units has shrunk significantly. Today, there are standard measuring units used throughout the world.

15.2.2 The meter

The **meter**, which is the international standard length unit, has an interesting history: It was originally defined as $1/10,000,000$ of the distance between the North Pole and the Equator. Since this definition is impractical, the meter was redefined in 1799 as the length of a particular platinum bar (this bar is displayed in a museum in Paris). This new definition was also imprecise, as the length of a solid body is affected by temperature changes. In 1869, the meter was redefined as the length of a bar less sensitive to environmental changes. In 1960, it was defined once again as $1,650,763.73$ wavelengths of the orange-red emission line in the electromagnetic spectrum of the krypton-86 atom in vacuum. In 1983, the meter was redefined (one last time?) as the distance travelled by light in vacuum in $1/299,792,458$ seconds. These frequent redefinitions did not significantly alter the length of a meter.

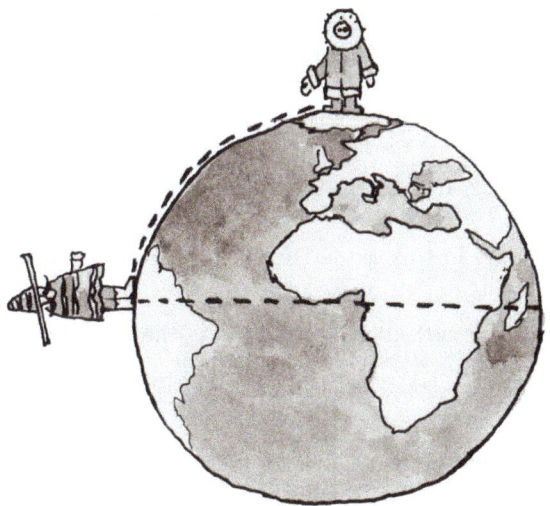

15.2.3 Using multiple measuring units

Suppose we measure the length of a boat and find out that the boat is 7 meters long. This means that if we concatenate 7 sticks, each one meter long, this chain of sticks will reach both ends of the boat. Suppose we used a different length unit, for example a foot, which is prevalent in most English-speaking countries. One foot is slightly less than a third of a meter. Obviously, the length of the boat does not change when we use a different measuring unit. It is only the numerical value of the measurement that changes. Measured in feet, the length of the boat is about 23 feet.

The connection between the measuring unit and the numerical value of the measurement is a consequence of the properties of division. The numerical outcome of the measurement is the quotient of the measured segment by the measuring unit. The laws of variation for division (Chapter 8, p. 107) imply that the reduction of the divisor—in the present case, the divisor is the measuring unit—results in the quotient increasing by the same factor. In simpler words, the shorter is the measuring unit, the larger is the numerical outcome of the measurement. When we switch from meters to feet, we reduce the divisor by a factor of about 3, and as a result, the numerical value of the measurement increases by the same factor.

15.2.4 Systems of measuring units

Suppose that there existed a unique measuring unit, say, the meter. We would then express the distance between Boston and San Francisco as about 4,300,000 meters, whereas the thickness of a hair would be expressed as about 0.0001 meters. Manipulating both very large and very small numbers is awkward, and more important, obstructs our number sense. In order to express lengths, such as the distance between cities, more conveniently we should use a measuring unit that is much longer than a meter. Likewise, in order to express lengths, such as the thickness of a hair, more conveniently we should use a measuring unit that is much shorter than a meter.

A **system of measuring units** is a set of measuring units with two characteristics:

(1) This set can be used to measure conveniently lengths over a wide spectrum of magnitudes.
(2) The various units in this set satisfy simple conversion rules.

The standard system of length-measuring units is the **metric system**. The units in this system are either products of a meter or parts of a meter, the factors being powers of ten.

Metric systems are prevalent not only in length measurement, but also in the measurement of mass, area and volume. The naming convention in metric systems is to attach numerical prefixes to a chosen base unit. In length measurement, the base unit is a meter. A hundredth of a meter is called a centimeter (cm), a thousandth of a meter is called a millimeter (mm), and thousand meters are called a kilometer (km).

The numerical prefixes centi-, milli-, and kilo- are derived from Latin. A table of metric prefixes along with their numerical value is displayed below:

15.2.5 Length-measuring instruments

Thus far, we have described the principles underlying length measurement. We will now address practical aspects of measurement. As a stimulating example, imagine that you want to measure the length of a stick about 100 cm long, and you use thin pieces of wood 1 cm long for measuring units. Using quotative division, we would have to take about a hundred of those measuring units, arrange them side by side next to the stick, until we obtain a chain of measuring units that extends between both ends of the stick. If that's not tiresome enough, we would also need to count how many measuring units were used.

In order to avoid such inefficient work, **length-measuring instruments** were invented. One of the most common length-measuring instruments is the **ruler**:

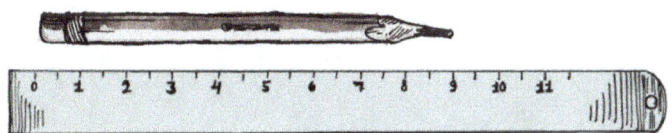

From a mathematical standpoint, a ruler is a line segment that can be placed conveniently alongside measured objects. The ruler is endowed with a line **gauge** that consists of a set of marks placed at fixed intervals. Some of the marks are labeled by numbers, with the first mark usually labeled 0. In rulers that use metric units, it is customary to enumerate marks at intervals of 1 cm. In rulers that use English units, it is customary to enumerate marks at intervals of 1 inch.

Consider the above figure which displays a metric ruler. The bottom end of the pencil is aligned with the mark labeled 0, and the tip of the pencil is aligned with the mark labeled 9. This means that the segment that connects the pencil's ends contains 9 segments of 1 cm. Hence, we conclude that the pencil is 9 cm long. The use of a measuring device endowed with a line gauge provides us with a ready-to-use chain of measuring units. The numbers on the gauge help us determine how many units there are between the two ends of a segment.

Standard rulers are useful for measuring lengths between 1 cm and 30 cm. Measuring tapes, which are long, flexible, retractable rulers, can be used to measure lengths up to several meters. Dedicated length-measuring instruments exist to measure lengths above several meters or below the scale of 1 cm.

15.3 The length of curves

Now that we understand what the length of a segment is (i.e., we know how to measure it), we can generalize the concept of length measurement to arbitrary curves.

15.3.1 The length of a broken line

Broken lines are chains of segments (Chapter 14, p. 178). If we sum up the edges of the broken line (in the sense of segment addition, as described in p. 179), we obtain a segment, whose length we know how to measure. This is precisely how we define the length of a broken line. In practice, there is no need to add edges. Instead, we can measure the length of each edge separately, thus obtaining a set of numerical values, which we then add together.

15.3.2 The length of more complicated curves

The length of a broken line is a very intuitive concept once we understand what the length of a segment is. The length of a curve (e.g., a circle) is a much less intuitive concept. In this section, we provide a rough sketch of how to define the length of a curve. A formal definition requires tools that are not learned until college.

Consider the following figure:

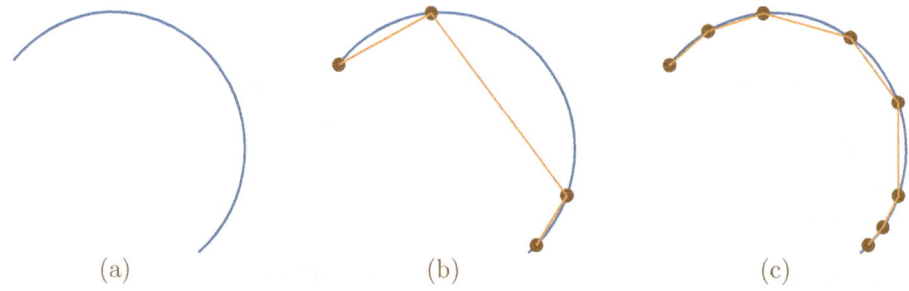

Figure (a) displays a curve. To evaluate its length we **approximate** the curve by a broken line by selecting several points on the curve and connecting them by segments. We then measure the length of the broken line (we already know how to do this), and this length approximates the length of the curve.

In Figure (b), we approximate the curve by a broken line by selecting four points on the curve. The length of this broken line is our first approximation of the length of the curve.

A quick glimpse tells us that a broken line comprising three segments is a very poor approximation to the given curve. In Figure (c), we **refine** the approximating broken line by appending four points to the four points used in Figure (b). We now have a broken line comprising seven segments. Its length, which is larger than the length of the broken line in Figure (b) (can you explain why?), is a finer approximation of the length of the curve.

If we keep refining the broken lines, using each time a set of vertices that is denser than its predecessor, we get a sequence of lengths, which approximate the length of the curve. This sequence of lengths keeps increasing, but at a slower rate. If we keep refining the broken line ad infinitum, we will discover that the sequence of approximating lengths approaches a **limit**. This limit is the length of the curve.

Even though the length of a curve is a complicated concept, the approximation procedure described above is not difficult to understand. Children often evaluate the length of a curve by chaining together paperclips, and then counting the total length of those paperclips. The approximation of a curve by a chain of paperclips is a concrete realization of its approximation by a broken line.

15.3.3 Infinitely long curves

The method outlined above for estimating the length of a curve can sometimes yield unexpected findings. One such finding was discovered by British physicist Lewis Fry Richardson (1881–1953) in the early 20th century. Richardson wanted to measure the length of the British coastline. In order to do that, he used the approximation technique described above.

Using a map of Great Britain, Richardson first approximated the British coastline by a closed broken line whose edges are 200 km long (Figure (a) below). By measuring the length of this broken line, he obtained a first approximation of the length of the British coastline. Next, he refined his estimate by approximating the coastline by a closed broken line whose edges are 100 km long (Figure (b) below). As expected, this refinement yielded a larger estimate for the length of the coastline. Next, he repeated the same procedure using 50 km long segments (Figure (c) below), and so on, reducing the length of the segments by a factor of 2 each time. As expected, the shorter the segments he used, the closer his broken lines reflected the shape of the British coastline. Much to his surprise, as the segments he used became shorter, the length of the broken lines increased in a manner that seemed unbounded. In other words, the sequence of approximating lengths kept growing to infinity. To his utter surprise, Richardson was forced to admit that the length of the British coastline is infinite.

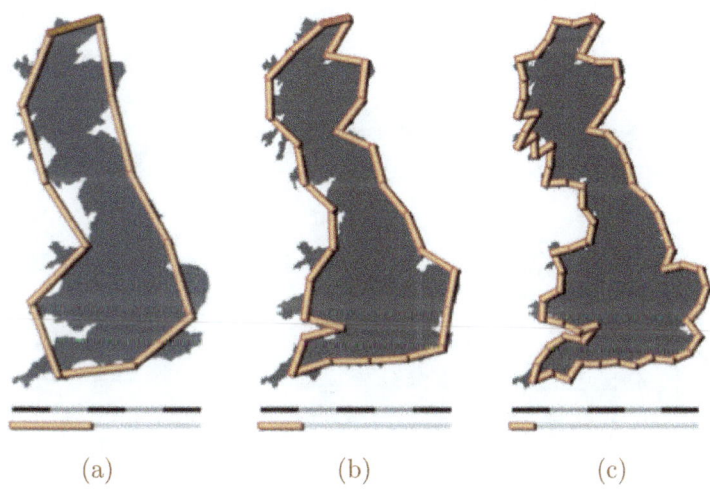

(a)   (b)   (c)

It was only several decades later that mathematicians established the existence of strange curves for which each part has an infinite length. Such curves are now known as **fractals**, which is a term coined by French physicist Benoît Mandelbrot (1924–2010).

Mathematical problems

Problem 15.1. In your own words, explain the connection between measurement and quotative division.

Problem 15.2. Convert the following English measuring units into meters: 1 inch, 1 foot, 1 yard, 1 mile.

Problem 15.3. Make a freehand drawing of a non-straight curve and estimate its length in two different ways:

(a) Place a string alongside the curve and then use a ruler to measure the length of the string.
(b) Approximate the curve by a broken line and measure the length of the broken line.

Which way seems more accurate? Why?

Problem 15.4. Estimate, without measuring, the height of the top doorpost. Estimate by how much your estimate can be off. Verify your estimate by measurement, and see if the exact height is within the expected range.

Chapter 16

Angles

This chapter introduces a new family of geometric figures—**angles**. Angles are fundamental in geometry for the same reason that segments are: both are building blocks for more complex figures. Angles also have a measurable attribute—they can be assigned a **size**. In this chapter, we get acquainted with angles and their measurement.

16.1 What is an angle?

Angle is a term used in everyday life. "The left striker scored a goal from a difficult angle", say the sportscasters, and the soccer fans understand what they mean. If we were to ask children which one of the two airplanes in the figure below is taking off, they would surely point to the one on the left, and explain that the plane is "at an angle".

If, however, we were to ask children (or adults) to define what an angle is, they might discover that they don't really know.

Angles constitute a family of geometric figures. They are planar figures, which means that they lie in a plane (and, consequently, can be drawn on flat paper). The standard definition of an angle is the following:

Definition 16.1. An **angle** is a geometric figure formed by two rays whose endpoints coincide.

The figure below displays two rays, AB and AC, whose endpoints coincide. By Definition 16.1, the two rays form an angle. The rays are called the **sides** of the angle. Their common endpoint is called the **vertex** of the angle.

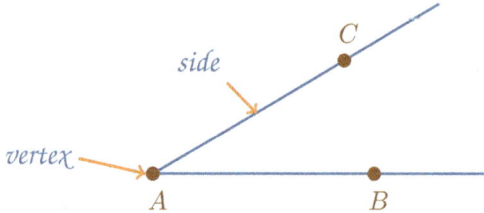

A ray is defined uniquely by two points. Two rays with a common endpoint are defined uniquely by three points: the common endpoint and two other points, one on each ray. Let A be the common endpoint of two rays and let B and C be two points, one on each ray. We denote the angle formed by the rays AB and AC by either

$$\sphericalangle BAC \quad \text{or} \quad \sphericalangle CAB.$$

By convention, the middle letter designates the vertex of the angle.

Two rays that intersect at a given point divide the plane into two domains. One domain is called the **interior** of the angle and the other is called the **exterior**. In a given angle, there is no way to tell which part of the plane is the interior and which part is the exterior unless you explicitly distinguish between them. In illustrations, it is customary to mark the interior of an angle by a sector that connects the sides of the angle, as shown below:

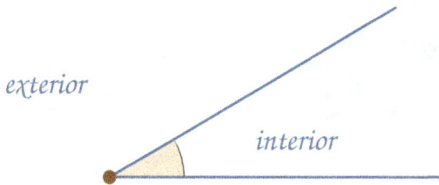

Angles can be given a slightly different definition, which is equivalent to Definition 16.1. This alternative definition will prove useful when we get to angle arithmetic:

Definition 16.2. A **wedge** is a part of a plane that is bounded by two rays whose endpoints coincide. An **angle** is the boundary of a wedge.

The figure below shows a part of a wedge (like a plane, a wedge extends to infinity). The boundary of the wedge comprises two rays that intersect at their endpoints, i.e., it is an angle. One advantage of Definition 16.2 is that it provides a natural distinction between the interior and the exterior of an angle: the wedge enclosed by the angle is the interior.

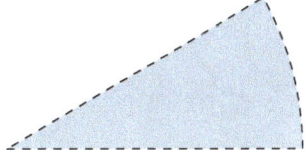

 Comments

(1) Two intersecting lines divide the plane into four wedges, and thus form four angles:

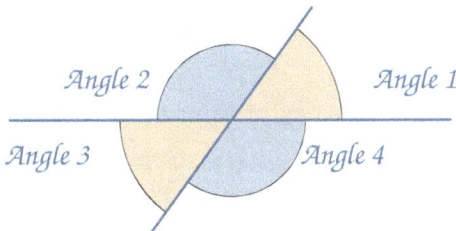

(2) The angle formed by two segments that intersect at a point is defined to be the angle formed by the rays that extend those segments.

16.2 Angle arithmetic

In Chapter 14, we introduced a class of geometric figures—segments—and proceeded to define an order relation and arithmetic operations between segments. Similarly, we now define an order relation and arithmetic operations between angles. In particular, quotative division of angles lays the foundations for angle measurement.

16.2.1 Angle comparison

Like any pair of geometric figures, two angles are **congruent** if one can be placed on top of the other such that they coincide exactly. The notion of congruence enables us to define an order relation between angles:

(a) Two angles are equal if they are congruent.
(b) Angle a is smaller than Angle b if the wedge enclosed by Angle a is congruent to a subset of the wedge enclosed by Angle b.

We express the order relation between angles using the relation signs $\cong$, < and >.

The following figure displays two angles, ∡a and ∡b:

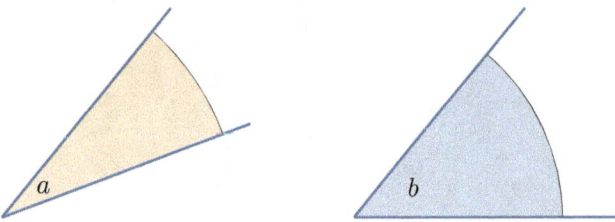

If we translate Wedge a such that its vertex coincides with the vertex of Wedge b, and we then rotate it until one of its sides coincides with a side of Wedge b,

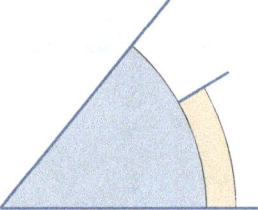

then we find that Wedge a is congruent to a subset of Wedge b, hence

$$\angle a < \angle b.$$

 A common misconception

An angle extends to infinity, even though we only sketch a finite part of it. Consider the two angles depicted below:

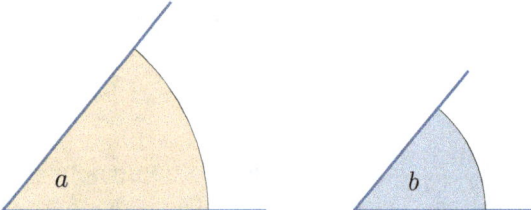

Children are inclined to think that Angle a is larger than Angle b, because the segments that form Angle a are longer than the segments that form Angle b. If, however, we superimpose the two angles, we will find out that they are congruent, i.e.,

$$\angle a \cong \angle b.$$

16.2.2 Angle addition

Angles can be added. The addition of angles, just like the addition of segments, is defined by concatenation. Consider the two angles depicted below:

Their sum is constructed as follows:

(a) Translate Wedge b such that its vertex coincides with the vertex of Wedge a.
(b) Rotate Wedge b until one of its sides coincides with a side of Wedge a (see figure below).

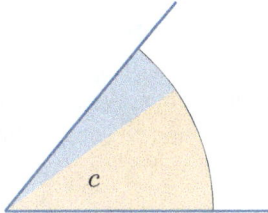

(c) Denote by Wedge c the compound wedge formed by Wedge a and Wedge b.
(d) The angle that encloses Wedge c is the sum of Angle a and Angle b, namely,

$$\sphericalangle c \cong \sphericalangle a + \sphericalangle b.$$

16.2.3 Angle subtraction

Subtraction can be viewed as the solution to an unknown-addend equation. Given two angles, $\sphericalangle a$ and $\sphericalangle b$, their difference is the angle, $\sphericalangle c$, whose sum with $\sphericalangle b$ yields $\sphericalangle a$. That is, the geometric expression

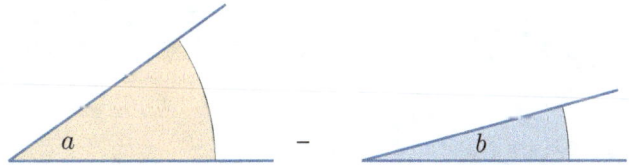

represents an angle that solves the unknown-addend equation:

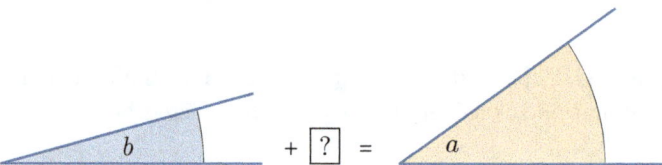

In practice, the difference between ∡a and ∡b is constructed as follows:

(a) Translate Wedge b such that its vertex coincides with the vertex of Wedge a.
(b) Rotate Wedge b such that it remains enclosed in Wedge a, and one of its sides coincides with a side of Wedge a (see figure below).

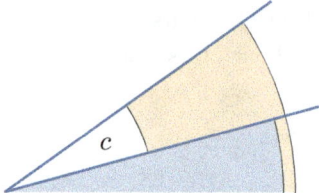

(c) Denote by Wedge c the part of Wedge a that is not covered by Wedge b.
(d) The boundary of Wedge c is the sought difference, namely,
$$\angle c \cong \angle a - \angle b.$$

16.2.4 Angle multiplication

Since we can add angles, the **repeated addition** of angles makes sense. In other words, we can multiply angles by numbers: the multiplier is a number, the multiplicand is an angle, and the product is an angle.

The following figure illustrates the product of 3 times a given angle:

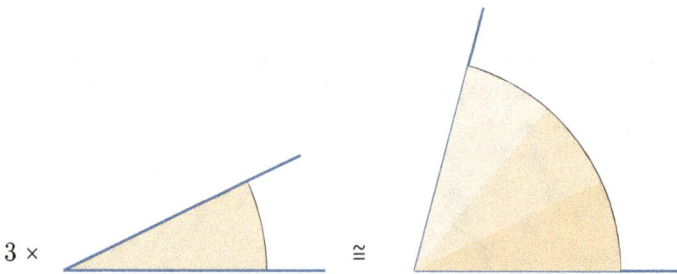

16.2.5 Angle division

If we can multiply angles, then we can also divide angles. Recall that division is an operation whose outcome can be interpreted as the solution to an unknown-factor equation. In this context, there is a fundamental distinction between partitive division and quotative division (cf. segment division on p. 181).

In **partitive division**, we know the multiplier (i.e., the number of portions), we know the product (i.e., the total amount), and the unknown is the multiplicand (i.e., the size of each portion). In the context of partitive angle division, the dividend is an angle, the divisor is a number and the quotient is an angle.

Consider the geometric expression:

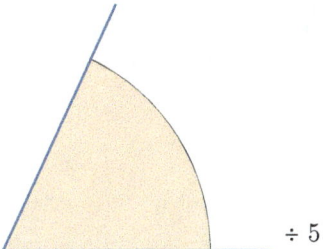

The quotient is an angle that, multiplied by 5, yields the dividend. Written as an unknown-factor equation:

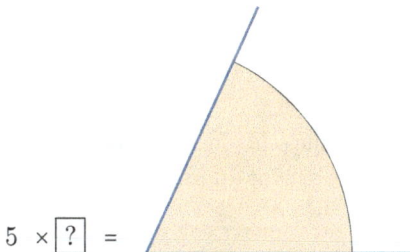

In **quotative division**, we know the multiplicand (i.e., the size of each portion), we know the product (i.e., the total amount), and the unknown is the multiplier (i.e., the number of portions). In the context of quotative angle division, the dividend is an angle, the divisor is an angle too and the quotient is a number.

Consider the geometric expression:

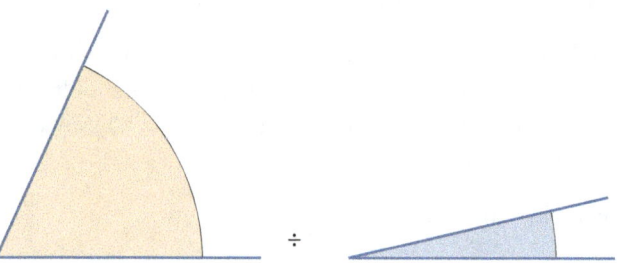

The quotient is the number of times that the divisor is contained in the dividend, or equivalently, the number whose product with the divisor yields the dividend. Written as an unknown-factor equation:

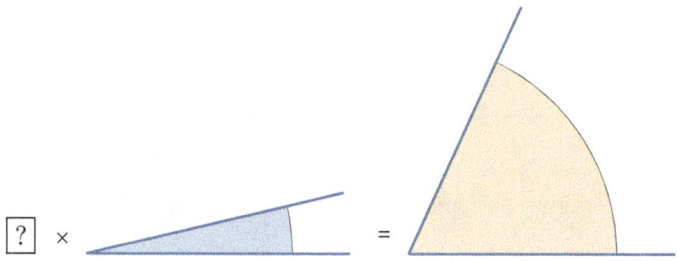

16.3 Angle measurement

Angle measurement follows the same scheme as the measurement of any other measurable attribute. When we measure the size of an angle, we are equipped with the following:

(a) An angle whose size we are measuring.
(b) An angle that serves as a **measuring unit**.
(c) A practical way of counting how many times the measuring unit is contained in the angle whose size we are measuring.

The outcome of the measurement is the number obtained in the last step.

16.3.1 The degree

There are standard measuring units for angles. The most common measuring unit is the **degree**. It has been in use for thousands of years, and can be traced back to ancient Indian manuscripts.

The degree is denotes by the symbol °. An angle of 1° is depicted below:

An angle of 1° is defined as follows: it takes 360 wedges of 1° to cover the entire plane. The number 360 turns out to be a recurrent number in ancient civilizations. It was considered a "convenient" number because it has many divisors, among which are all the natural numbers up to 10, except for 7.

Thus, when we say that the size of an angle is 60°, we mean that it contains 60 angles of 1°, or equivalently, that it can be obtained by joining together 60 angles of 1°.

16.3.2 Types of angles

Angles are classified into types according to their size:

(a) A **straight angle** is an angle that encloses a half-plane (Angle a below). Alternatively, we may say that the sides of a straight angle form a line. Since the wedge enclosed by a straight angle is half of the wedge that encloses an angle of 360°, the size of a straight angle is 180°.

(b) A **right angle** is half of a straight angle (Angle b below). It follows that the size of a right angle is 90°. Two lines or segments that form a right angle are said to be **perpendicular** to each other.

(c) An **acute angle** is an angle that is smaller than a right angle (Angle c below). That is, the size of an acute angle is less than 90°.

(d) An **obtuse angle** is an angle that is larger than a right angle and smaller than a straight angle (Angle d below). That is, the size of an acute angle is greater than 90° and smaller than 180°.

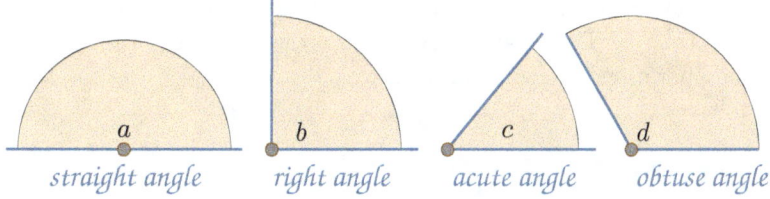

16.3.3 The protractor

Just as we do not measure length by actually adding up segments of 1cm, we also do not measure angles by actually adding up angles of 1°. Angles are measured using a measuring instrument—a **protractor**.

A protractor is a sort of "angular ruler" with a straight base topped with a semicircle.

The center of the base is marked, and the semicircle has two gauges of equidistant marks. Both gauges range between 0° and 180°; one gauge ascending clockwise and the other gauge ascending counter-clockwise.

To measure the size of an angle, we place the center of the base right above the angle's vertex. We then rotate the protractor until one of the angle's sides coincides with the protractor's base, and the other side intersects the semicircle (see figure below). The point of intersection indicates the size of the angle.

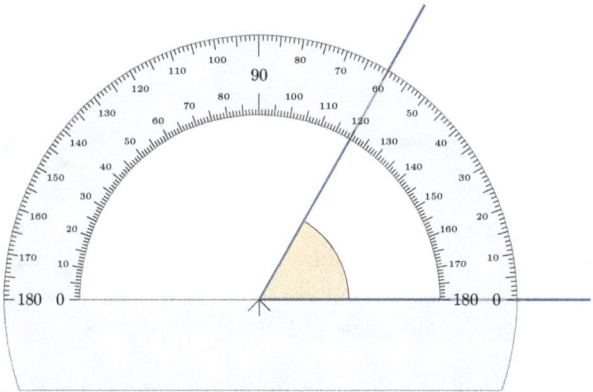

In this figure, the point of intersection is 60° according one gauge and 120° according the other gauge. The relevant gauge is the one whose zero mark is intersected by the other side of the angle. Thus, the size of the angle in the present case is 60°.

 **Activity: Drawing angles**

Ask children to draw angles of various sizes (for example: 30°, 45°, 70° and 135°) using a ruler and a protractor. This is an instructive exercise, as it requires the planning and execution of an operation inverse to measuring.

16.4 Parallel lines

Two lines can intersect at most at one point; we proved this theorem on p. 175. Thus, given two lines, there are only two possible scenarios:

(a) The lines intersect at a single point.
(b) The lines do not intersect at all.

In this section, we consider the second scenario where two lines do not intersect.

In three-dimensional space, it is not hard to imagine two lines that do not intersect. Draw any line on the room's floor and draw any other line on the room's ceiling (of course, you only draw segments and you have to imagine those segments extending to infinity in both directions). It is intuitively clear that those lines do not intersect.

In general, two lines, one on the floor and one on the ceiling, do not lie on a joint plane. Non-intersecting lines are of particular interest when the two lines lie on the same plane. In this context, Greek geometers defined the following term:

Definition 16.3. Lines that lie on the same plane and do not intersect are called **parallel**.

Consider the two lines that extend the two segments depicted below:

The lines get closer the further we move to the left, until they eventually intersect. Thus, these lines are not parallel.

Consider now a line a and an arbitrary point B that is not on that line:

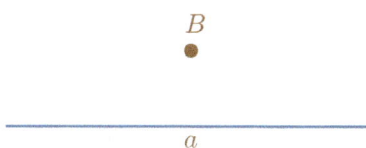

There are infinitely many lines on that plane that pass through point B, a few of which are depicted below:

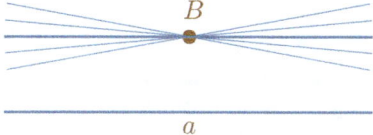

Some of those lines intersect Line a on the right whereas some of those lines intersect Line a on the left. Intuitively, there has to exist a critical line that passes through Point B and does not intersect Line a.

The existence of a parallel line that passes through Point B seemed obvious to Euclid. He was convinced that he would be able to prove its existence using the postulates that he already had at his disposal. After many years of failed attempts, he was forced to admit that the existence of a parallel line that passes through Point B could not be proven. He finally formulated a new postulate, known as **Euclid's fifth postulate**:

> **Postulate 16.1.** Given a line a and a point B that does not lie on Line a, there is a unique line that passes through Point B and is parallel to Line a.

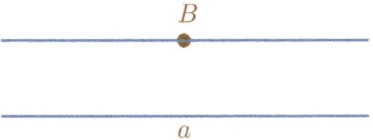

Euclid's fifth postulate enabled Greek geometers to prove many new theorems. An important consequence of the fifth postulate is depicted below: Given two parallel lines, a and b, and a line, c that intersects both lines, the so-called **alternating angles** (see figure below) are equal.

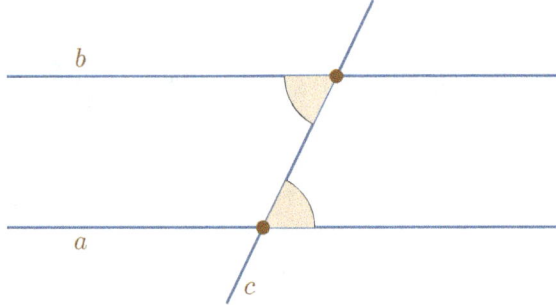

This fact may seem intuitive, based on symmetry considerations, but it could not be proven without Euclid's fifth postulate.

In particular, it follows that if Lines a and b are parallel, and Line c is perpendicular to Line a, then it is necessarily perpendicular to Line b as well (see figure below):

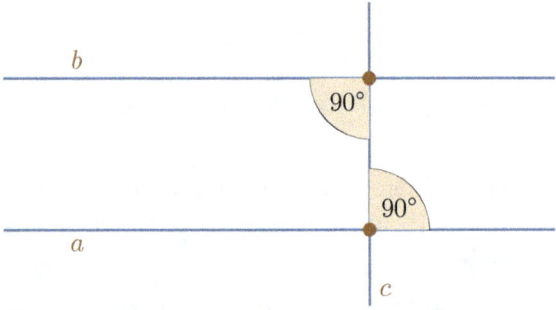

Finally, a notion of parallelism is defined for segments:

> **Definition 16.4.** Two segments are said to be parallel if the lines extending from those segments are parallel.

For example, the two segments depicted below are parallel:

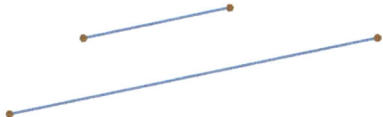

How can we tell that those segments are indeed parallel? In principle, we would have to extend those segment indefinitely and verify that they do not intersect. This is, of course, impractical. A practical test for parallelism is based on angle measurement. Intersect those segments with a transversal line and verify, using a protractor, that the alternating angles are equal. In elementary school, we often rely on visual perception and informal reasoning. For example, children might argue that these segments are parallel because they maintain a fixed distance from each other.

 Please note

Children often think that for two segments to be parallel they must be either horizontal or vertical, and they must have equal length. These are misconceptions that should be rebutted through the encounter of numerous counter-examples.

Mathematical problems

Problem 16.1. Use precise mathematical language to explain what the sportscasters mean when they say "that goal was scored from a difficult angle".

Problem 16.2. What is the size of the unknown angles in the following figure?

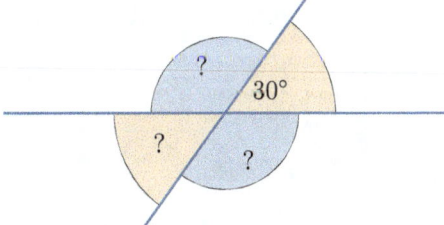

Justify your answer.

Problem 16.3. Draw a broken line according to the following specification: The line is made up of five segments measuring 3, 5, 12, 2 and 5 cm. The angle between the first segment and the second segment is 120°. The angle between the second segment and the third segment is 100°. The angle between the third segment and the fourth segment is 60°. The angle between the fourth segment and the fifth segment is 150°.

Problem 16.4. Draw a triangle and measure the size of its angles. What is the sum of the three angles? Repeat the same measurement with another triangle of different shape. What do you observe?

Problem 16.5. Draw two parallel segments on a blank sheet of paper, with a 30°-60°-90° set square at your disposal (a triangular ruler often used in technical drawing).

Pedagogical problems

Problem 16.6. Draw a line (preferably neither horizontal nor vertical) and a point that is not on that line (see the figure below):

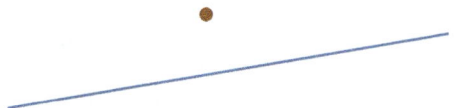

Ask the children to measure the distance of that point to the line. Let them to explain their measuring procedure using informal language. Ask them to explain the meaning of the distance of a point from a line.

Chapter 17

Polygons

In Chapter 14, we introduced a family of geometric figures called **broken lines**. A broken line is a sequence of points, together with the segments that connect each point to its successor. A broken line is called **closed** if it includes, in addition, the segment that connects the first vertex to the last vertex.

Plane geometry is concerned with geometric figures that lie in a plane. Broken lines are not necessarily **planar figures**. Take any sequence of points in space that do not lie on a plane and connect them with segments—the resulting figure is a non-planar broken line. This chapter is concerned with a class of geometric figures that are planar, closed, broken lines.

17.1 What is a polygon?

We start by defining what a polygon is:

Definition 17.1. A **polygon** is a planar, closed broken line.

The figure below displays a polygon. There are five points, ordered as follows: A, B, C, D and E. A segment connects the first point, A, to its successor, B. Another segment connects the second point, B, to its successor, C, and so on. Finally, a segment connects the last point, E, to the first point, A.

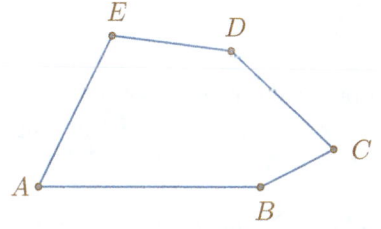

Next, we introduce some terminology associated with polygons:

(a) The points that delineate the broken line are called the **vertices** of the polygon.
(b) The segments that connect the vertices are called **edges**, or **sides**.
(c) Two vertices that are connected by an edge are called **adjacent vertices**.
(d) Two sides that have a vertex in common are called **adjacent sides**.
(e) A segment that connects two non-adjacent vertices is called a **diagonal**.
(f) A polygon partitions the plane into two domains: an **interior** and an **exterior**. The interior is called a **polygonal domain**.
(g) Every two adjacent sides form an angle. The interior of the angle is always facing the interior of the polygon.

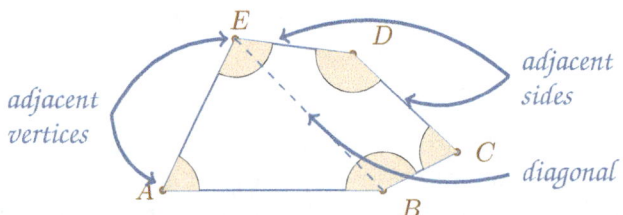

A polygonal domain is a two-dimensional figure that is bounded by straight edges. Polygonal domains are ubiquitous in geometry for various reasons. First, they are relatively simple to construct and to combine, which makes them useful in engineering and in architecture. Second, measurable attributes of polygonal domains (e.g., **perimeter** and **area**) can be calculated from formulas. This makes polygonal domains ideal shapes if, for example, we were asked to partition land into properties of equal area.

The most basic classification of polygons is by the number of sides (which equals also the number of vertices and the number of angles). A three-sided polygon is called a **triangle**. A four-sided polygon is called a **quadrilateral**. For polygons with five edges and more, there is a fixed naming scheme: the name of the polygon comprises a Greek numerical prefix that corresponds to the number of edges, followed by the suffix -gon.

 Polygons in elementary school

The study of polygons occupies a significant part of the elementary school curriculum. Specifically, the study of polygons focuses on the following topics:

(1) **Primary classification**: Naming polygons according to the number of sides:

(2) **Secondary classification**: Polygons, and, in particular, triangles and quadrilaterals, are classified into secondary classes (e.g., rectangles are a class of quadrilaterals). The secondary classification of polygons is according to properties of their sides, properties of their angles, and combinations of both. Thus, children learn to identify types of triangles and quadrilaterals, and name them appropriately. This learning sometimes bears some resemblance to a visit to the zoo, where children learn that "this is a giraffe" and "this is an elephant".

(3) **Defining properties and derived properties**: The set of properties that a polygon has to fulfill in order to belong *by definition* to a certain class of figures are called the **defining properties** of that class of figures. For example, for a polygon to belong to the class of rhombuses it must have four sides and those sides have to be of equal length. Polygons that belong to a certain class may satisfy additional properties pertinent to figures in that class. Those properties are referred to as **derived properties**. For example, a derived property of parallelograms is that opposite angles are equal. It is a property that holds for all parallelograms, but it is not part of the standard definition of a parallelogram.

The distinction between defining properties and derived properties is often a matter of convention. For example, we could have defined equilateral triangles as triangles in which all three angles are equal. In such a case, the equality of the three sides would have been a derived property. The distinction between defining and derived properties is difficult to grasp in the early school years, and I do not recommend delving into it too deeply in elementary school.

(4) **Measurable attributes**: Polygons have measurable attributes: the length of their sides, the length of their diagonals, the length of their perimeter, the size of their angles, and the area of their interior. Children learn to measure lengths and angles in polygons using measuring instruments. In addition, they learn to calculate the perimeter and the area of certain polygons from given data.

17.2 Triangles

A triangle is a three-sided polygon (in the context of defining versus derived properties, the defining property of a triangle is by the number of sides). The smallest number of sides that a polygon can have is three, hence triangles are the simplest type of polygons.

17.2.1 The sum of the angles in a triangle

The most important derived property of triangles is that the sum of the size of their angles equals 180°.

Consider the following activity:

 **Activity: Angles in a triangle**

(a) Ask each child in the classroom to draw a triangle.
(b) Present the various triangles to the class, and point out that they are all different.
(c) Ask children to measure the three angles in their own triangle using a protractor.
(d) Write the measurement data on the board, and in each case, calculate the sum of the size of the three angles.
(e) Show that in each case the sum is either 180°, or at least close to it. The deviation from 180° is due to measurement inaccuracies.

The above activity might leave a child perplexed: why should the sum of the three angles of a triangle always be 180°? In other words, how are the three angles in a triangles related to a straight angle?

While a formal argument has to await at least until middle school, an intuitive explanation can be provided in elementary school. Consider an arbitrary triangle, in which each of the three angles has been colored differently:

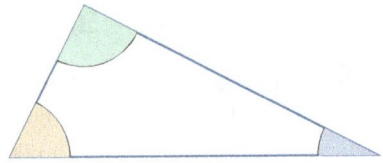

If we take three replicas of the *same* triangle, and then translate and rotate them, we can assemble them as follows:

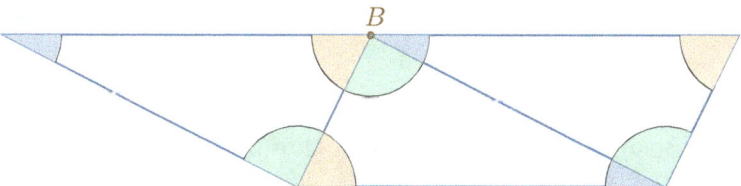

Point B is the vertex of three angles. On the one hand, these three angles are the three angles of the original triangle. On the other hand, the sum of these three angles is a straight angle, which is precisely what we wanted to show.

17.2.2 Classification of triangles according to their sides

Now we will introduce the secondary classification of triangles. In this section, the defining properties of the secondary classifications will be according to properties of sides.

> **Definition 17.2.** A triangle in which all three sides have equal length is called an **equilateral triangle**.

The figure below displays a triangle. If you measure its sides, you will find out that they are all 4 cm long, hence, this triangle is equilateral.

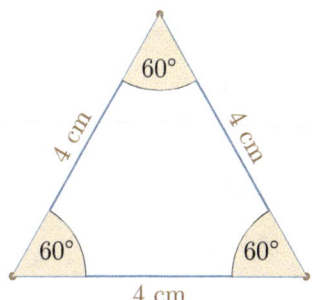

Equilateral triangles constitute the most symmetric type of triangles. When we say that a geometric figure displays a **symmetry**, we mean that there exists certain rotations and/or reflection that leave the figure unchanged. The symmetry of equilateral triangles implies that not only their sides, but also their angles are equal (the equality of the angles is a derived property of the class of equilateral triangles). Since the angles in a triangle add up to 180°, it follows that each angle in an equilateral triangle is of 60°.

The next class of triangles is characterized by a joint property of only two sides:

Definition 17.3. A triangle that has two sides of equal length is called an **isosceles triangle**.

The figure below displays a triangle whose vertices are A, B and C (in short, $\triangle ABC$). If you measure its sides, you will discover that the sides AC and BC have equal length, hence, this triangle is isosceles.

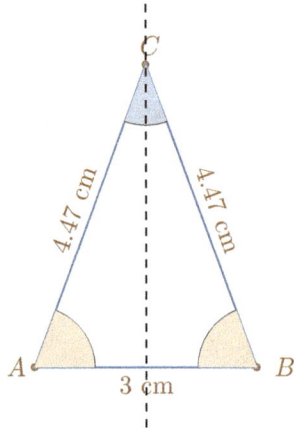

The third side (which may or may not be equal to the two equal sides) is called the **base** of the triangle. The two angles adjacent to the base are called **base angles**. The angle opposite the base is called the **head angle**.

Isosceles triangles exhibit the following symmetry: if we reflect the triangle about the line that bisects the head angle (the dashed line in the figure) we recover the exact same triangle. Differently stated, the head angle bisector partitions the triangle into two congruent triangles. In particular, it follows that the base angles of an isosceles triangle are equal.

17.2.3 Classification of triangles according to their angles

Thus far, we have seen two types of triangles that are characterized by properties of their sides. Other types of triangles are characterized by properties of their angles.

Definition 17.4. A triangle in which all three angles are acute is called an **acute triangle**. A triangle that has a right angle is called a **right triangle**. A triangle that has an obtuse angle is called an **obtuse triangle**.

In a right triangle, the side opposite the right angle is called the **hypotenuse**. The other two sides are called the **catheti** (singular: **cathetus**) of the triangle. Please note that since the angles in a triangle add up to 180°, there can be at most one right angle or one obtuse angle.

The three types of triangles are displayed below:

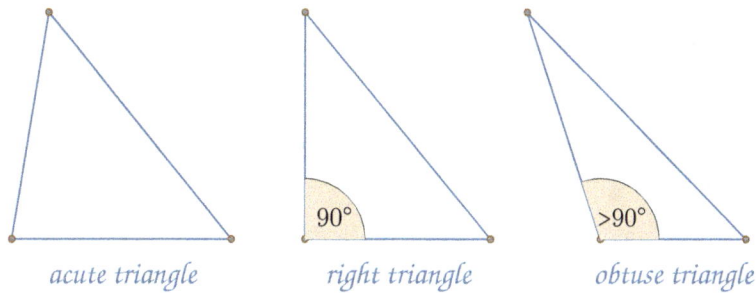

acute triangle   *right triangle*   *obtuse triangle*

17.2.4 Inclusion relations between types of triangles

Consider the following questions:

(1) Are all equilateral triangles also isosceles?
(2) Are all isosceles triangles also equilateral?
(3) Does there exist an equilateral triangle that is also isosceles?
(4) Does there exist an isosceles triangle that is also equilateral?

Children (and adults alike) tend to get confused by these sort of questions, which are ubiquitous throughout the study of geometry. Our goal in this section is to offer a systematic approach to questions like these.

The setting through which we should examine the above four questions is **set theory**. Polygons constitute a set of geometric figures. Triangles are a subset of the set of polygons. This implies that although every triangle is a polygon, there are polygons that are not triangles (e.g. pentagons). Equilateral triangles and isosceles triangles are both subsets of the set of triangles. As such, they are also subsets of the set of polygons.

When we ask whether "all equilateral triangles are also isosceles" we ask in fact whether the set of equilateral triangles is a subset of the set of isosceles triangles. Conversely, when we ask whether "all isosceles triangles are also equilateral" we are asking whether, in fact, the set of isosceles triangles is a subset of the set of equilateral triangles. When we ask whether "there exists an equilateral triangle that is also isosceles" we ask in fact whether the set of equilateral triangles and the set of isosceles triangles are not disjoint.

The answer to all four questions follows from the defining properties of those classes of triangles:

(a) A triangle is a three-sided polygon.
(b) An equilateral triangle is one in which *all three* sides are of equal length.
(c) An isosceles triangle is one in which *there are two* sides of equal length.

Since in an equilateral triangle all three sides are of equal length, there are at least two sides of equal length. Thus, every equilateral triangle qualifies as an isosceles triangle (and in set-theoretical terms, the set of equilateral triangles is a subset of the set of isosceles triangles). The converse is not true. There are isosceles triangles that are not equilateral (i.e., triangles that have two equal sides, and a third side is not equal to the first two). The set of isosceles triangles is not a subset of the set of the equilateral triangles. On the other hand, there are isosceles triangles that are also equilateral (all equilateral triangles are isosceles triangles).

Inclusion relations among sets of geometric figures can be illustrated by means of a **Venn diagram**:

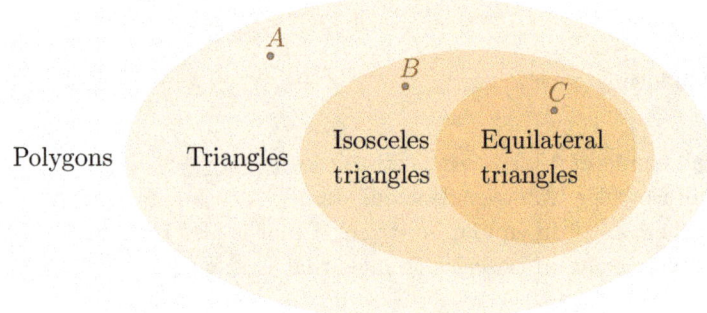

In this diagram, the largest ellipse represents the set of all polygons. Figuratively speaking, every possible polygon is represented by a point in this ellipse. Within the set of all polygons, there exists a subset that contains all the triangles. The set of all triangles is represented by the set of points in the second largest ellipse. Within the set of all triangles, there exists a subset that contains all the isosceles triangles. The isosceles triangles are represented by an ellipse contained in the ellipse representing the set of all triangles. Finally, within the set of all isosceles triangles, there is a subset containing all the equilateral triangles. This fact is illustrated by an ellipse that represents the set of equilateral triangles, which is contained in the ellipse representing the set of all isosceles triangles.

Consider Point A in the diagram. It represents a polygon that belongs to the set of triangles, but does not belong to the set of isosceles triangles (and, consequently, does not belong to the set of equilateral triangles). Point B represent an element in the set of isosceles triangles that does not belong to the set equilateral triangles. Point C represents an element in the set of equilateral triangles. Since the set of equilateral triangles is contained in the set of isosceles triangles, Point C also represents an element in the set of isosceles triangles.

17.2.5 Congruent triangles and constructions

Like any pair of geometric figures, two triangles are congruent if they only differ by a rigid motion (i.e., one can be obtained from the other through a combination of translations, rotations and reflections).

Two triangles are congruent if, and only if, their respective sides and their respective angles are equal. Consider the two triangles depicted below:

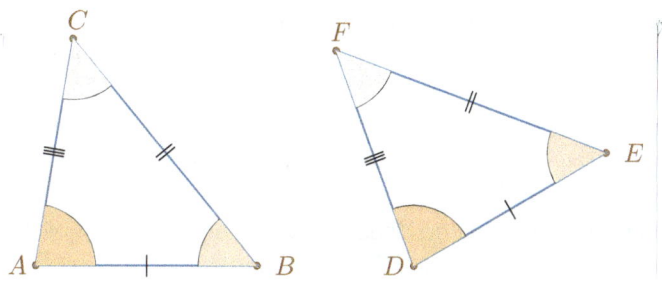

Triangles $\triangle ABC$ and $\triangle DEF$ are congruent. When we specify a congruence relation between polygons, we have to specify which vertex in one polygon corresponds to which vertex in the other polygon. In the present case, Point A corresponds to Point D, Point B corresponds to Point E, and Point C corresponds to Point F. Thus, the correspondence between sides and angles in both triangles is as follows:

(a) Side AB is equal to Side DE.
(b) Side AC is equal to Side DF.
(c) Side BC is equal to Side EF.
(d) Angle $\sphericalangle BAC$ is equal to Angle $\sphericalangle EDF$.
(e) Angle $\sphericalangle ABC$ is equal to Angle $\sphericalangle DEF$.
(f) Angle $\sphericalangle ACB$ is equal to Angle $\sphericalangle DFE$.

Given two triangles, $\triangle ABC$ and $\triangle DEF$, if all six conditions (a)–(f) are satisfied, then the two triangles must be congruent. This implies that in constructing a triangle $\triangle ABC$, given the length of all its sides and the size of all its angles, for example:

$$AB = 4\,\text{cm} \qquad AC = 3\,\text{cm} \qquad BC = 2\,\text{cm}$$
$$\sphericalangle BAC = 28.9° \qquad \sphericalangle ABC = 46.6° \qquad \sphericalangle ACB = 104.5°,$$

there exists only one such triangle—every two triangles with the same dimensions are congruent.

A fundamental question in Euclidean geometry is what the minimal conditions are that uniquely determine a triangle. For example, it is clear that there are infinitely many triangles $\triangle ABC$ that only satisfy the requirement $AB = 2$ cm. But what about triangles $\triangle ABC$ that satisfy two conditions, say, $AB = 2$ cm and $AC = 3$ cm? Is there only one such triangle? Are there many?

It turns out that a unique triangle can be determined by just specifying the length of one side, along with two other attributes: either the length of another side or the size of an angle.

Specifically:

(1) A unique triangle is determined by specifying the lengths of two of its sides and the size of the angle enclosed by those two sides (the so-called Side-Angle-Side, or SAS criterion).
(2) A unique triangle is determined by specifying the length of one of its sides and the size of the two angles adjacent to that side (the so-called Angle-Side-Angle, or ASA criterion).
(3) A unique triangle is determined by specifying the length of its three sides (the so-called Side-Side-Side, or SSS criterion).

The SAS, ASA and SSS criteria are studied within the framework of deductive geometry in middle and high school. Then, the attributes that determine a unique triangle are used to prove new theorems, and children rarely stop to think why these criteria hold. Elementary school geometry provides an exceptional opportunity to understand these uniqueness criteria through a concrete hands-on experience.

 **Activity: Drawing triangles**

The following activity can be performed with a group of children, each equipped with a ruler and a protractor.

(1) Ask each child in class to draw a triangle with one of its sides 2 cm long. Display all the results. They will surely discover that they differ from each other (see figures below). It can now be deduced that the length of one side does not determine a triangle uniquely.

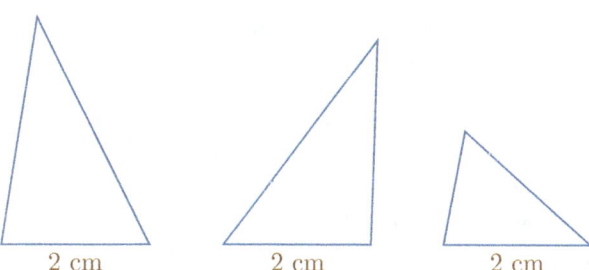

(2) Ask each child in class to draw a triangle with one side 2 cm long and another side 3 cm long. Display all the results. This exercise will show that not all of the triangles are identical (see figures below). We can therefore deduce that the lengths of two sides does not determine a unique triangle.

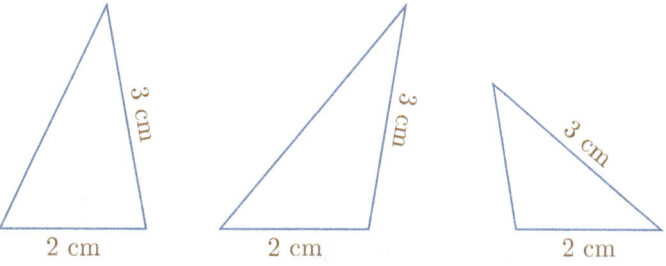

(3) Ask each child in class to draw a triangle with its sides 2 cm, 3 cm, and 4 cm long. Please note that children may have a hard time doing it. One possible strategy is to ask them to construct a material triangle from sticks of desired lengths.

Another strategy is displayed below.

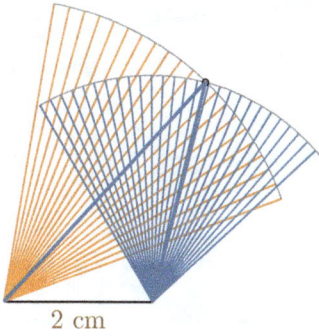

First, draw a side of 2 cm. From one of its ends, draw an arc of radius 4 cm. From the other end, draw an arc of radius 3 cm. The intersection of those two arcs is the third vertex of the triangle. Display all the results. This exercise will show that all the triangles are congruent. We can now deduce that the lengths of all three sides determines a unique triangle.

(4) Ask each child in class to draw a triangle with one side 3 cm long, another side 4 cm long, and the angle enclosed between those sides 50°. Display all the results. This exercise will show that all the triangles are congruent (see figure below). We can now deduce that the lengths of two sides and size of the angle enclosed between them determines a unique triangle.

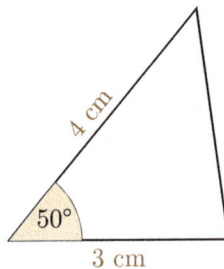

17.3 Quadrilaterals

A quadrilateral is a four-sided polygon (in the context of defining versus derived properties, the defining property of a quadrilateral is by the number of its sides). Quadrilaterals are sub-classified according to properties of their sides, angles, and combinations of both. Since quadrilaterals have more sides and angles than triangles, they have more subclasses than triangles.

17.3.1 The sum of the angles in a quadrilateral

We already know that the sum of the angles in a triangle is 180°. This fact can be used to deduce that the sum of the angles in a quadrilateral is also fixed.

Take an arbitrary quadrilateral, such as the one depicted below:

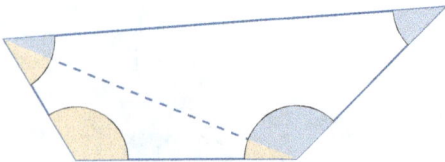

The broken line is a diagonal. It partitions the quadrilateral into two triangles. Since the sum of the angles in each triangle is 180°, and since the angles of the quadrilateral comprise the angles of both triangles, it follows that the sum of the angles in a quadrilateral is twice 180°, that is 360°. Since the same argument holds for any quadrilateral, we can therefore deduce an important derived property of quadrilaterals: the sum of their angles is 360°.

17.3.2 Squares

Definition 17.5. A **square** is a quadrilateral that is both **equilateral** and **equiangular**. That is, its four sides are equal and its four angles are equal.

Since the sum of the angles in any quadrilateral is 360°, it follows that each angle is 90°, i.e. it is a right angle.

Squares are the most symmetric quadrilaterals; they are an instance of **regular polygons** (see below). In addition to their defining properties, squares also satisfy the following derived properties:

(a) Opposite sides are parallel.
(b) The two diagonals are equal.
(c) The two diagonals are perpendicular to each other.

17.3.3 Rectangles

Definition 17.6. A **rectangle** is an **equiangular** quadrilateral. That is, its four angles are equal.

Since the angles of a rectangle are equal, each is 90°, i.e., a right angle (in fact, the word *rectangle* means right angle).

Rectangles satisfy the following derived properties:

(a) Opposite sides are parallel.
(b) Opposite sides are equal.
(c) The two diagonals are equal.

Please note that every square is a rectangle, or stated differently, the set of squares is a subset of the set of rectangles. The opposite is not true: not every rectangle is a square.

17.3.4 Rhombuses

Definition 17.7. A **rhombus** (also called a **diamond**) is an **equilateral** quadrilateral. That is, its four sides are equal.

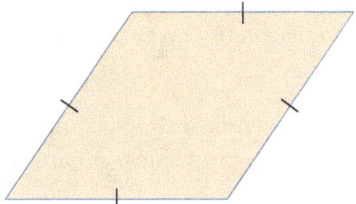

Rhombuses satisfy the following derived properties:

(a) Opposite sides are parallel.
(b) Opposite angles are equal.
(c) Adjacent angles sum up to 180°.
(d) The two diagonals are perpendicular to each other.

Please note that every square is a rhombus, or differently stated, the set of squares is a subset of the set of rhombuses. The opposite is not true: not every rhombus is a square. On the other hand, if a quadrilateral is both a rhombus (equal sides) and a rectangle (equal angles), then it is a square.

17.3.5 Parallelograms

Definition 17.8. A **parallelogram** is a quadrilateral whose opposite sides are parallel.

Parallelograms satisfy the following derived properties:

(a) Opposite sides are equal.
(b) Opposite angles are equal.
(c) Adjacent angles sum up to 180°.

Please note that every square, rectangle, or rhombus is a parallelogram (because all three satisfy the derived property of opposite sides being parallel).

17.3.6 Trapezoids

Definition 17.9. A **trapezoid** is a quadrilateral that has (at least) one pair of opposite sides that are parallel.

Please note that every parallelogram is also a trapezoid, or stated differently, the set of parallelograms is a subset of the set of trapezoids. The opposite is not true: there are rectangles in which only one pair of opposite sides are parallel.

17.3.7 Kites

Definition 17.10. A **kite** (also called a **deltoid**) is a quadrilateral that has two pairs of adjacent sides that are equal in length.

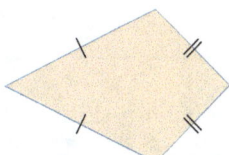

A kite can be viewed as two isosceles triangles that have a joint base.

Kites satisfy the following derived properties:

(a) The angles between the non-equal sides are equal.
(b) The two diagonals are perpendicular to each other.

Please note that every square or rhombus is a kite, but the opposite is not true. There are kites that are not rhombuses.

17.4 General polygons

17.4.1 The sum of the angles

We have already seen that in all triangles the sum of the angles is 180°, whereas in all rectangles the sum of the angles is 360°. Is it generally true that the sum of the angles in a polygon depends only on the number of sides? That is, is it true that the sum of the angles is the same for all pentagons, it is the same for all hexagons, and so on?

To understand the rule that relates the number of sides to the sum of the angles, consider a pentagon.

Choose a vertex, it does not matter which, and draw all the diagonals that emanate from that vertex. Since every vertex in a pentagon has two non-adjacent vertices, there are two such diagonals. The two diagonals partition the pentagon into three triangles. Since the angles of the pentagon comprise exactly the angles of the three triangles, it follows that the sum of the angles in a pentagon is three times the sum of the angles in a triangle, namely, $3 \times 180° = 540°$.

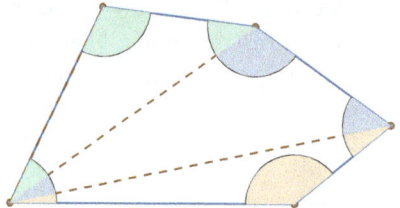

This argument can be generalized for an n-sided polygon. By drawing all the diagonals that emanate from a vertex, the polygon can be partitioned into $n - 2$ triangles. Since the angles of the polygon comprise exactly the angles of $n - 2$ triangles, it follows that

> The sum of the angles in an n-sided polygon = $(n - 2) \times 180°$.

17.4.2 The number of diagonals

The number of diagonals in a polygon only depends on the number of sides. The following activity offers a guided exploration of the relation between the number of sides and the number of diagonals.

 **Activity: Diagonals in a polygon**

(1) Start with triangles. Triangles are polygons with the minimal number of sides. Since every two vertices in a triangle are adjacent, there are no diagonals.

(2) Next consider quadrilaterals. Ask each child to draw a quadrilateral along with all its diagonals. Comparing result, children will realized that all quadrilaterals have exactly 2 diagonals.

(3) Move on to pentagons and hexagons. Ask each child to draw a pentagon and a hexagon along with with all their diagonals. This may require a more complex tracking that every two non-adjacent points are connected by a diagonal. Ask each child to report how many diagonals each shape has.

(4) By now, we can start to fill out a table that relates the number of sides to the number of diagonals:

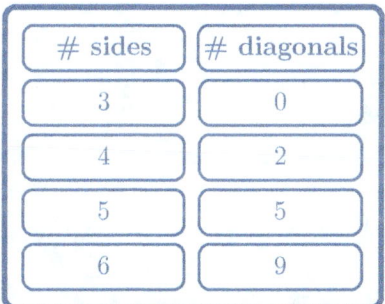

# sides	# diagonals
3	0
4	2
5	5
6	9

(5) Encourage the children to find the rule that connects between the number of sides and the number of diagonals. For example, ask them to predict the number of diagonals in a 7-sided polygon, or in a 100-sided polygon.

(6) By the end of the discussion, the following explanation can be given. Consider a 100-sided polygon. It has 100 vertices. Every vertex has 97 non-adjacent vertices, because we have to subtract from the 100 vertices the vertex itself and two adjacent vertices. Thus, there are 100×97 ways to draw an arrow from a vertex to a non-adjacent vertex. To find the number of diagonals, we have

to divide the latter expression by 2, because we counted every diagonal twice. Thus,

$$\text{The number of diagonals in a 100-sided polygon} = \frac{100 \times 97}{2}.$$

Replacing the number 100 by any number n, we obtain the following relation:

$$\text{The number of diagonals in an } n\text{-sided polygon} = \frac{n \times (n-3)}{2}.$$

(7) Finally, verify that this formula is indeed correct for triangles, quadrilaterals, pentagons and hexagons.

17.4.3 Regular polygons

A **regular polygon** is a polygon that is both equilateral (all its sides are equal) and equiangular (all its angles are equal). We have already seen two types of regular polygons: equilateral triangles and squares. The following figure depicts 3-sided to 8-sided regular polygons.

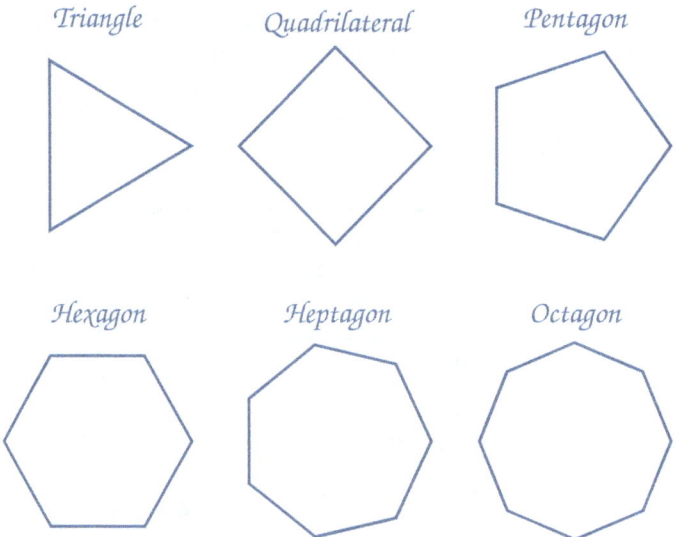

Greek geometers had a particular interest in geometric figures that can be constructed using only a **straightedge** (i.e., an unmarked ruler) and a **compass**. Regular triangles and quadrilaterals, for example, can be constructed quite easily with those instruments. Constructing a 17-sided regular polygon, on the other hand, is a hard task. In 1796, the German mathematician Carl Friedrich Gauss (1777–1855) succeeded. Certain regular polygons, however, seemed impossible to construct by a straightedge and a compass. It was only in 1837 that the French mathematician Pierre Wantzel (1814–1848) proved that there are regular polygons that cannot be constructed in this way.

The sequence of natural numbers n, for which it is possible to construct by a straightedge and a compass an n-sided regular polygons starts as follows:

$$3, 4, 5, 6, 8, 10, 12, 15, 16, 17, 20, 24, \ldots$$

Thus, Wantzel proved that polygons with 7, 9, 11, 13, 14, 18 or 19 sides cannot be constructed with a straightedge and a compass.

Mathematical problems

Problem 17.1. Draw a triangle that is both isosceles and obtuse.

Problem 17.2.

(a) Draw a triangle whose sides are 6 cm, 5 cm, and 3 cm long.
(b) Draw a triangle whose sides are 10 cm, 5 cm, and 3 cm long.
(c) If this is difficult, explain why.

Problem 17.3. Draw a triangle according to the following specification: the sides measure 5 cm and 6 cm, and the angle between them is 45°.

Problem 17.4. Draw a right triangle with the two catheti 3 cm and 4 cm long. Measure the length of the hypotenuse.

Problem 17.5. Explain why a quadrilateral that is both a rhombus and a rectangle is a square.

Problem 17.6. Draw two different rectangles with a perimeter of 10 cm. Draw two different rhombuses with a perimeter of 10 cm.

Problem 17.7. Can there be a quadrilateral with three sides of equal length and a fourth side of a different length?

Problem 17.8. Can there be a quadrilateral with three angles of equal size and a fourth angle of a different size?

Problem 17.9. Determine whether each of the following claims is true of false:

(a) There exists a rhombus that is also a rectangle.
(b) Every trapezoid is also a parallclogram.
(c) There exists a rhombus that is also a rectangle, and that is not a square.
(d) There exists a square that is also a trapezoid, and that is not a kite.

(e) There exists a kite that has a right angle and that is not a square.
(f) There exists a parallelogram that has a right angle and that is not a rectangle.
(g) There exists a rhombus that has a right angle and that is not a square.

Problem 17.10. To which type of quadrilaterals does the following figure belong?

Problem 17.11. Does there exist a pentagon with an angle larger than 180°? If so, please draw it. Does there exist a pentagon with two angles larger than 180°?

Problem 17.12. Draw a Venn diagram that expresses the inclusion relations between the types of quadrilaterals introduced in this chapter.

Problem 17.13. Draw a square. Which of the following shapes can be obtained by changing the location of just one of its vertices?

(a) A rectangle
(b) A parallelogram
(c) A trapezoid
(d) A kite
(e) An equilateral triangle
(f) An isosceles triangle

Problem 17.14. What is the size of each angle in a regular pentagon? In a regular hexagon? In the regular 100-sided polygon?

Chapter 18

Area

Area is a measurable attribute of two-dimensional figures (i.e., **surfaces**). Loosely speaking, the area of a surface is a measure of the space that this surface occupies. We have already met an important class of surfaces—**polygonal domains**—which are **planar surfaces**. Another type of planar surfaces are **discs** (the interior of **circles**). An example of a non-planar surface is a **sphere**. We can attribute an area to all these types of figures.

In Chapter 15, we studied a measurable attribute of one-dimensional figures, **length**. Most of that chapter was devoted to the length of segments. Only toward the end of Chapter 15, did we examine the length of non-straight curves, building upon the length of segments. A similar scenario occurs with the concept of area. First, we define the area of rectangles. Then, we use this concept to define the area of more complicate planar surfaces. Finally, the area of non-planar surfaces builds upon the area of planar surfaces.

The elementary school curriculum is concerned with the area of planar figures. The course of study exposes children to the concept of area, teaches them to calculate the area of a selected class of figures (rectangles, triangles and discs), and to estimate the area of more complex figures.

18.1 The area concept

Area is a measurable attribute of surfaces. In other words, every surface can be attributed a number—its area—which expresses the space that it occupies. In our general introduction to measurable attributes (Chapter 13, p. 169), we argued that the only way to define measurable attributes is by specifying how to measure them. We define the concept of area through the following principles:

- **DP1**: Area is an attribute of surfaces.
- **DP2**: The value of area is always a positive number.
- **DP3**: Two figures that are congruent have the same area.
- **DP4**: If a surface is partitioned into disjoint parts, then the area is the whole equals the sum of the areas of the parts.

(The prefix DP stands for *defining principle*.)

Comments

(1) Congruent figures have the same area. For two figures to have the same area, however, they do not need to be congruent. For example, a square whose side is 1 cm long and a rectangle whose sides are 2 cm by 0.5 cm are not congruent, but they have the same area.

(2) The concept of area is similar to the concept of length; in Defining Principles 1–4, if you replace *surface* by *curve* and *area* by *length*, you obtain the defining properties of length.

(3) Defining Principles 1–4 are not common to all measurable attributes. For example, it is not true that if we partition a body into disjoint parts, then the temperature of the whole equals the sum of the temperatures of the parts.

(4) Even though area is an attribute of surfaces, it is common to attribute a vanishing area to zero-dimensional figures (i.e., points) and to one-dimensional figures (i.e., curves).

18.2 Area comparison

The concept of area is best understood through the study of representative examples.

Consider the following two figures:

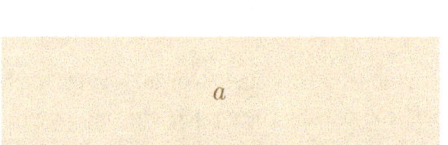

Which figure has the largest area?

Figures a is a rectangle and Figure b is square (which is a type of rectangle, p. 220). Since the two figures are not congruent, we cannot establish the equality of the areas based on Defining Principle 3—congruence. Instead, we may try to use Defining Principle 4, and partition the figures into parts.

Below, Figures a and b are partitioned into two parts each:

Figure a is partitioned into two rectangles, which we label a_1 and a_2. Figure b is partitioned into two rectangles, which we label b_1 and b_2. If we measured the sides of the four rectangles, a_1, a_2, b_1 and b_2, we would find out all four rectangles are congruent. According to Defining Principle 3, it follows that

$$\text{Area of } a_1 = \text{Area of } a_2 = \text{Area of } b_1 = \text{Area of } b_2.$$

By Defining Principle 4, it follows that

$$\text{Area of } a = \text{Area of } a_1 + \text{Area of } a_2$$
$$\text{Area of } b = \text{Area of } b_1 + \text{Area of } b_2.$$

The two sets of equalities imply that

$$\text{Area of } a = \text{Area of } b.$$

 Please note

We haven't really specified what area is, and yet, based on Defining Properties 1–4, we have been able to determine that two figures have equal area. Specifically, we **tiled** both figures with congruent figures, and found that both require the same number of tiles. At this point, we still do not assign area a numerical value.

Consider two other figures:

Which figure has the largest area?

The two figures are not congruent, hence we cannot compare them based on Defining Principle 3. Once again, we tile the two figures with congruent figures, this time triangles:

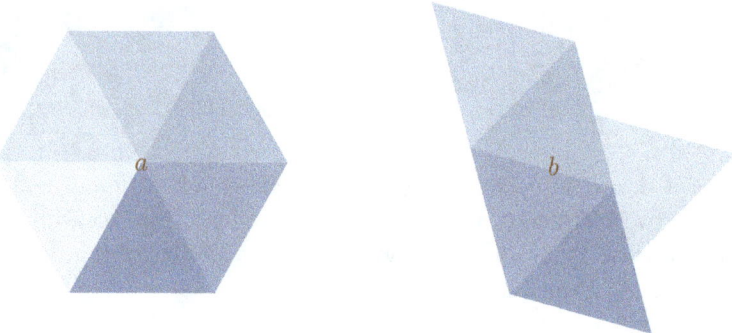

We don't know what the area of a triangle is, but we know that Figure a contains six such triangles, whereas Figure b contains five such triangles. It follows from Defining Principle 4 that

$$\text{Area of } a = 6 \times \text{Area of one triangle}$$
$$\text{Area of } b = 5 \times \text{Area of one triangle}.$$

By Defining Principle 2, the area of one triangle is positive. Since 6 times a positive number is greater than 5 times that number, it follows that

$$\text{Area of } a > \text{Area of } b.$$

The next example displays two new figures:

Which figure has the largest area?

This time the two figures are not regular enough to be tiled by congruent shapes. Instead, we place both figures on a grid of congruent squares:

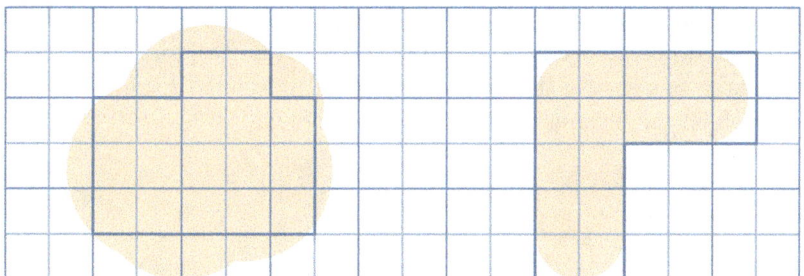

Figure a contains 17 squares, whereas Figure b is contained in 16 squares. Thus,

$$\text{Area of } a > 17 \times \text{Area of one square}$$

$$\text{Area of } b < 16 \times \text{Area of one square}.$$

By Defining Principle 2 the area of one square is positive, hence

$$\text{Area of } a > \text{Area of } b.$$

18.3 Area measurement

By now, we have introduced all the concepts needed to define area measurement. As in any other type of measurement, when we measure the area of a surface, we are equipped with the following:

(1) A surface whose area we are measuring.
(2) A surface which we call a **measuring unit**.
(3) A practical mean of counting how many times the measuring unit is contained in the surface whose area we are measuring.

The outcome of the measurement is the number obtained in the last step.

The standard units of area are squares. A square whose side is 1 cm long is a unit of area called a **square-centimeter** (sq-cm). A square whose side is 1 inch long is a unit of area called a **square-inch** (sq-in), and so on.

Thus, standard units of area are derived from standard units of length. This is the source of a very common misconception among children: the confusion of length and area. Length and area are two distinct measurable attributes. As such, they cannot be compared. It makes absolutely no sense to say that the length of a segment is greater than the area of a rectangle (see Problem 18.6)!

How do we measure the area of a surface? In principle, we have to select a measuring unit, say a sq-cm, tile the surface with that measuring unit, and count how many times the measuring unit is contained in the measured figure. In many cases, this course of action cannot be implemented. For example, you cannot tile a disc with squares. Thus, the concept of the measurement of area has to be refined beyond the tiling procedure.

In the next section, we will study the area of simple polygons—rectangles and triangles. More complicated polygons can be partitioned into triangles, and therefore their area can be inferred from a sum of the areas of triangles. In Volume 2, we will study the area of a disc, which is a significantly more complex concept.

18.4 The area of polygons

18.4.1 Rectangles

Consider a rectangle whose sides are 5 cm and 3 cm long:

What is its area?

When length is given in centimeters, it is natural to choose for unit area a square-centimeter. If we partition the sides of the rectangle into segments of 1 cm, we can partition the whole rectangle into an array of squares:

The side of each square is 1 cm, hence its area is, by definition, 1 sq-cm. Thus, we tiled the rectangle with an array of 5×3 squares, each of 1 sq-cm. It follows that

$$\text{Area of rectangle } = 5 \times 3 \text{ sq-cm} = 15 \text{ sq-cm}.$$

If we had a rectangle whose sides were 9 cm and 7 cm long, we would follow the same steps, and tile it with an array of 9×7 square-centimeters. If the sides of the rectangles were 6 m and 4 m long, we would tile it with an array of 6×4 squares whose sides are 1 m. Thus, we would determine that its area is 24 square-meters (sq-m).

Summarizing:

> The area of a rectangle whose sides are a-by-b length units is $a \times b$ units of area, where a unit of area is a square whose sides are one length unit.

 Please note

People often interpret the above conclusion as the area being the product of the sides. This is inaccurate. Instead, one should say that the number of units of area contained in a rectangles is the product of the number of length units contained in a pair of adjacent sides.

18.4.2 Triangles

Consider the following triangle:

What is its area? In principle, we should try to tile it with unit squares. It is clear, however, that a triangle cannot be tiled with squares.

Here is an ingenious trick. Enclose the triangle in a rectangle, and draw a line that connects a vertex of the triangle to the opposite side, intersecting at a right angle.

A segment that connects a vertex of a triangle to the opposite side, and intersects it at a right angle is called a **height** of a triangle. The side that intersects the height is often called a **base**. Thus, Segment AB is the height of the triangle that emanates from vertex A. The side on which lies Point B is the corresponding base.

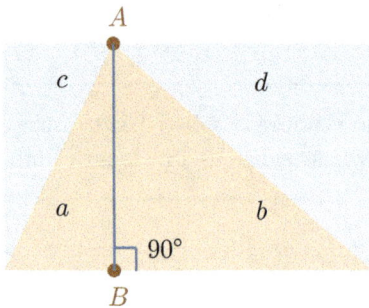

Segment AB partitions the rectangle into two smaller rectangles. Each of the smaller rectangles is divided into two right triangles, one of which is part of the original triangle, and one of which is not. We label those four triangles a, b, c and d.

Now,
$$\text{Area of rectangle} = \text{Area of } a + \text{Area of } b + \text{Area of } c + \text{Area of } d,$$
whereas
$$\text{Area of triangle} = \text{Area of } a + \text{Area of } b.$$
Triangle a and Triangle c are congruent, and Triangle b and Triangle d are congruent. It follows from Defining Principle 3 that
$$\text{Area of } a = \text{Area of } c \quad \text{and} \quad \text{Area of } b = \text{Area of } d,$$
and consequently, the area of the rectangle is twice the area of the triangle. Equivalently, the area of the triangle is half the area of the rectangle.

Thus, to determine the area of the triangle we need to know the area of the rectangle that contains that triangle. The area of the rectangle in unit areas equals the product of its sides in unit lengths. One of the rectangle's sides is equal to the height AB, whereas the other side coincides with the corresponding base.

This leads us to the following conclusion:

> The area of a triangle (in square units) is equal to half of the product of a height of that triangle times the length of the corresponding base (both given in same length units).

 Three different heights

Please note that a triangle has three heights, and therefore can be enclosed by three different rectangles, as shown below:

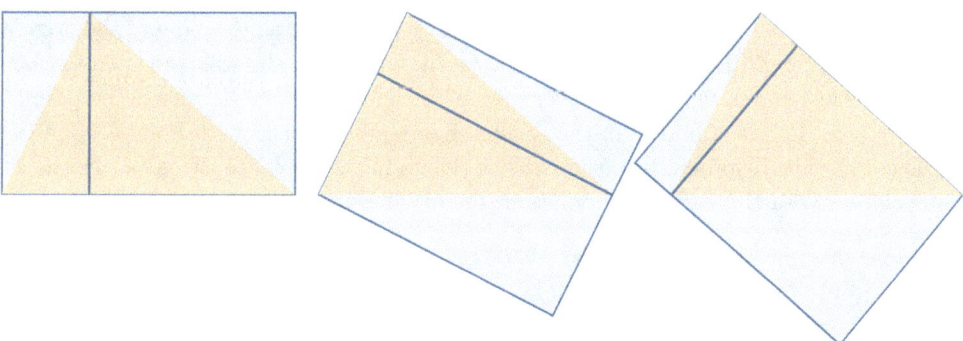

Those three enclosing rectangles are not congruent, but they have equal area, because their areas are twice the area of the same triangle.

18.4.3 General polygons

Once we know how to calculate the area of any triangle by measuring two lengths—a height and a base—in principle, we can calculate the area of any polygon. Take, for example, an arbitrary quadrilateral. By tracing a diagonal, we can partition the quadrilateral into two triangles. The area of the quadrilateral is the sum of the areas of the two triangles. Similarly, any polygon can be partitioned into triangles by tracing all the diagonals that emanate from one of its vertices.

18.5 Area and scaling

Suppose you draw a geometric figure, say, a triangle, and insert it into a magnifying photocopy machine. In a photocopy machine, you can select the level of magnification. If you select a factor-2 magnification, you get something of that sort:

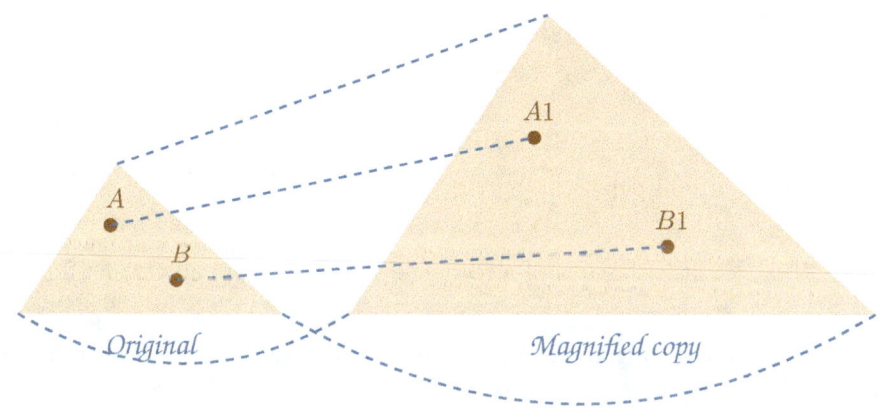

What does a factor-2 magnification mean? If you measured the sides of both triangles, you would find out that the sides of the copy are twice as long as the sides of the original. In fact, *every length* has been magnified by a factor of 2 in the following sense: there exists an exact matching between every point in the original figure and every point in the copy (e.g., Points A and B in the original figure are matched with Points A_1 and B_1 in the copy). The distances between pairs of points in the copy are twice as large as the distances between the respective points in the original figure.

How does magnification affect area? If the original figure were a rectangle, then all of its sides would be magnified by 2, and as a result, its area, whose value is the product of its sides, would be multiplied by 4. Likewise, if the original figure were a triangle, both its base and its height would be magnified by 2, and as a result its area would be multiplied by 4. If we had magnified a shape by a factor of 3, its area would have grown by a factor of 3×3, i.e., 9.

This leads us to the following conclusion:

> When a two-dimensional figure—whichever one—is magnified by a factor a, its area is magnified by a factor $a \times a$.

Mathematical problems

Problem 18.1. Estimate the area in sq-cm of the following shape:

Use either a ruler or grid paper.

Problem 18.2. What is the area of a rectangle whose sides are 30 cm and 2 m long?

Problem 18.3. In the Kingdom of Clover, the standard unit of area is a right triangle whose catheti are 1 cm long. This unit is called a *triangular-centimeter* (abbr. tr-cm). What is the area of a rectangle whose sides are 6 cm and 5 cm long in tr-cm?

Problem 18.4. What are the areas (in square units) of the three squares depicted below? Base your answer on arguments available to elementary school students.

Problem 18.5. In many photocopy machines, the standard enlargement ratio is 1.414, and the standard reduction ratio is 0.707. Explain why these particular numbers were chosen.

Pedagogical problems

Problem 18.6. Teacher Signe presented a rectangle whose sides are 3 cm and 2 cm long to her class:

Colin claimed that the perimeter of this rectangle, which is 10 cm, is larger than its area, which is 6 sq-cm.

  (a) What is the misconception underlying Colin's claim?
  (b) What would you reply to Colin?
  (c) How would Colin's claim be affected if we measured lengths in units of millimeters and area, consequently, in units of square-millimeters?

Index

acute angle, 201
add to, 32, 47
addend, 26, 32
addition, 31, 155
    associative property, 36, 40, 45
    commutative property, 35, 45
    identity property, 39
    laws of variation, 40
    of angles, 197
    of segments, 179
    properties, 34
    within 20, 43
addition machine, 25, 36, 37
addition table, 45, 156
adjacent sides, 208
adjacent vertices, 208
algebraic equation, 35
algebraic expression, 35, 61
algorithm, 121
alternating angles, 204
angle, 168, 193, 195
    addition, 197
    arithmetic, 195
    comparison, 195
    division, 199
    measurement, 200
    multiplication, 198
    subtraction, 197
append, 32
area, 168, 208, 227
arithmetic expression, 27
arithmetic operations, 25
augend, 26, 32

axiom, 163

backward counting, 64
base, 233
base angle, 212
base-5 system, 151
betweenness, 176
binary operation, 36
blank number line, 65
broken line, 178, 189, 207
bundling, 17

cardinality, 4, 12
carry, 135
cathetus, 212
chunking, 114
circle, 227
combinatorics, 88
comparison
    angles, 195
    length, 183
    numbers, 7
    sets, 6
compass, 224
complementation, 52, 56
complexity, 126
concatenation, 184
congruence, 167, 195, 227
counting, 3
counting on, 42
counting-at-a-glance, 5
crossing ten, 43
curve, 172, 183, 189

decimal fractions, 124
decimal system, 15
decimal unit, 17, 121
deductive geometry, 163, 164
defining property, 209
degree, 200
deltoid, 221
derived property, 209
diagonal, 208, 223
difference, 53
digit, 15, 18
digits
   Hindu-Arabic, 18, 154
disc, 227
disjoint, 165
dividend, 26, 99
divisibility, 69
divisibility rule, 74
division, 97, 160
   adjoint equation, 108
   by five, 116
   by one, 112
   by ten, 115
   by zero, 110
   distributive property, 104
   evaluation, 113
   laws of variation, 106, 116, 187
   not associative, 103
   not commutative, 103
   of angles, 199
   of segments, 181
   of zero, 109
   partitive, 70
   quotative, 70
   sign, 99
divisor, 26, 99

edge, 178, 208
empty set, 5, 39, 165
equation, 28
error control, 76
estimation, 10
estrangement, 151
Euclid

fifth postulate, 204
Euclidean geometry, 163
evaluation problem, 28
even, 69
exact matching, 3, 236
exterior, 208

factor, 26, 81
forward counting, 42, 45, 64, 160
fractal, 192
fraction division, 102
fractions, 1
fundamental concepts, 163

gauge, 189
geometric figure, 166, 171
grouping, 17

head angle, 212
height, 233
heptagon, 209
hexagon, 209
hundred table, 21
hypotenuse, 212

identity element, 87
if you can count, you can...
   add, 41
   divide, 113
   multiply, 91
   subtract, 63
infinity, 9
interior, 208
intersection, 165

join together, 31

kite, 221

length, 168, 179, 183, 227
line, 172

mass, 168
matching

bijective, 2
exact, 3
measurable attribute, 168, 183, 227
measurements, 168
measuring unit, 169, 183, 185, 200, 231
meter, 186
minuend, 26, 53
modeling, 28
multiplicand, 26, 80
multiplication, 79, 159
   associative property, 84
   by zero, 87
   commutative property, 82
   distributive property, 86, 94
   evaluation, 91
   identity element, 87
   laws of variation, 87
   of angles, 198
   of segments, 181
multiplication table, 92, 160
multiplier, 26, 80

natural numbers, 1
negative numbers, 1, 66
non-standard representation, 125, 157
number line, 9, 46, 54, 67
number sequence, 4
numeral system
   China, 23
   Egypt, 22
   Rome, 23
numerals, 4

obtuse angle, 201
octagon, 209
odd, 69
one dimensional, 172
operand, 26
operator, 26
order relation, 6

parallel, 203
parallelogram, 220
parity, 69, 158

addition table, 73
arithmetic, 72
multiplication table, 73
partitive division, 97, 98, 100, 181, 199
pentagon, 209
perimeter, 208
place holder, 18, 124
place value, 18, 122, 154
planar figure, 172, 207
plane, 171
plane geometry, 172, 207
point, 165
polygon, 207
polygonal domain, 208, 227
portion, 97
postulate, 163, 172
predecessor, 8
product, 81
proof, 164
proof by contradiction, 175
proof by example, 75, 84
proportional comparison, 80
proportional reasoning, 106
protractor, 201, 205, 210, 216

quadrilateral, 209, 218
quinary system, 151
quota, 97
quotative division, 97, 99, 100, 168, 182, 183, 199
quotient, 99

rational numbers, 1
ray, 177
real numbers, 1
rectangle, 219
   area, 232
rectangular array, 82, 91
regrouping, 44, 121, 125, 129, 132, 139, 141, 157
   multiple, 143
regular polygon, 219, 224
relation signs, 8
repeated addition, 79, 88, 113, 198

repeated subtraction, 114
rhombus, 220
right angle, 201
ruler, 189, 216

scaling, 235
segment, 177, 183
  addition, 179
  arithmetic, 179
  comparison, 179
  division, 181
  multiplication, 181
  subtraction, 180
set, 165
  infinite, 12
set theory, 11, 165, 213
sharing, 97
side, 194, 208
sign
  addition, 32
  division, 99
  equal, 28
  multiplication, 80
  subtraction, 53
skip count, 113
skip counting, 93
space, 165, 171
sphere, 227
square, 219
square-centimeter, 231
square-inch, 231
standard measuring unit, 186
straight angle, 201
straightedge, 224
subset, 165
subtract
  to-ten, 65
subtraction, 49, 158
  adjoint equation, 62
  evaluation, 63
  French algorithm, 146
  laws of variation, 58
  not associative, 58
  not commutative, 57
  of angles, 197
  of segments, 180
  properties, 57
  zero, 63
subtrahend, 26, 53
successor, 8
surface, 171, 227

table
  addition, 45
  hundred, 21
  multiplication, 92
take apart, 51, 56
take away, 50, 56
temperature, 168
theorem, 164
three dimensional, 165
trailing zeroes, 124
transitivity, 6, 11
trapezoid, 221
tree diagram, 89
triangle, 209, 210
  acute, 212
  area, 233
  classification, 211
  congruence, 215
  equilateral, 211
  isosceles, 212
  obtuse, 212
  right, 212
two dimensional, 171

unary representation, 126
unary system, 17
union, 165
unknown-addend equation, 50–52, 158, 180, 197
unknown-factor equation, 99, 109, 112, 115, 160, 181, 199
unknowns, 53

validation, 66
Venn diagram, 214
vertex, 178, 194, 208

vertical addition, 130, 134
vertical subtraction, 140
volume, 168

wedge, 194
word problems, 28, 34, 55, 88, 100

zero, 5, 39, 63, 87, 109, 124

Common Core Index

1.MD.1, 184
1.MD.2, 185
1.NBT.2, 18
1.OA.1, 43, 65
1.OA.3, 45
1.OA.4, 50–52
1.OA.5, 41, 63
1.OA.6, 43, 65
1.OA.7, 28, 64
2.G.1, 208, 210, 218
2.G.2, 232
2.MD.1, 188
2.MD.2, 186
2.MD.6, 9
2.NBT.1, 18
2.NBT.3, 123
2.NBT.6, 129, 139
2.NBT.7, 129, 132, 139, 141
2.NBT.9, 132, 141
2.OA.2, 43, 65
2.OA.3, 76
3.G.1, 219–221
3.MD.5, 172, 231
3.MD.6, 231

3.MD.7, 232
3.NBT.2, 130, 134, 140, 141, 144
3.OA.1, 79, 81
3.OA.2, 100
3.OA.5, 93, 94
3.OA.6, 98, 99, 181, 182, 199
3.OA.7, 115, 116
4.AO.1, 80
4.MD.5, 194, 200
4.MD.6, 201
4.MD.7, 197
4.NBT.2, 123
4.NBT.4, 130, 134, 140, 141

K.CC.2, 42
K.CC.3, 5
K.CC.4, 4, 8
K.CC.6, 6
K.MD.2, 183
K.OA.1, 41, 63
K.OA.2, 41, 43, 63
K.OA.3, 43
K.OA.4, 43

www.ingramcontent.com/pod-product-compliance
Lightning Source LLC
Chambersburg PA
CBHW080612230426
43664CB00019B/2867